Transport interactions between gas and water in thin hydrophobic porous layers

Von der Fakultät Energie-, Verfahrens- und Biotechnik
der Universität Stuttgart zur Erlangung der Würde
eines Doktors der Ingenieurwissenschaften (Dr.-Ing.)
genehmigte Abhandlung

vorgelegt von
Stefan Dwenger
geboren in Reutlingen

Hauptberichter: Prof. Dr.-Ing. U. Nieken
Mitberichter: Prof. Dr.-Ing. J. Groß

Tag der mündlichen Prüfung:
15.12.2015

Institut für Chemische Verfahrenstechnik
der Universität Stuttgart
2016

Bibliografische Information der Deutschen Nationalbibliothek

Die Deutsche Nationalbibliothek verzeichnet diese Publikation in der
Deutschen Nationalbibliografie; detaillierte bibliografische Daten sind
im Internet über http://dnb.d-nb.de abrufbar.

D 93 (Diss. Universität Stuttgart)

ISBN 978-3-8325-4197-2

Logos Verlag Berlin GmbH
Comeniushof, Gubener Str. 47,
10243 Berlin
Tel.: +49 (0)30 42 85 10 90
Fax: +49 (0)30 42 85 10 92
INTERNET: http://www.logos-verlag.de

für Judith

Vorwort

Die vorliegende Arbeit entstand im Rahmen meiner Tätigkeit als wissenschaftlicher Mitarbeiter am Institut für Chemische Verfahrenstechnik der Universität Stuttgart sowie als Mitglied des internationalen Graduiertenkollegs NUPUS.

Besonderer Dank gilt Prof. Dr.-Ing. Ulrich Nieken, der als Institutsleiter und Betreuer zum Gelingen der Arbeit entscheidend beigetragen hat. Die gewährten Freiheiten und Möglichkeiten schufen Raum für eigenständiges und kreatives Arbeiten sowie Chancen zum Erlernen neuer Dinge. Prof. Dr.-Ing. Joachim Groß danke ich herzlich für die gewährte Gastfreundschaft an der TU Delft und für die Übernahme des Mitberichts.

Großer Dank geht an Manuel Huber, der jederzeit hilfsbereit und diskussionsfreudig (auch als Partner bei Fragestellungen des virtuellen Materialdesigns) zur Verfügung stand. Franz Keller, als erstem Bürokollegen und Kickerpartner, danke ich für viele spannende Kicker-Matches ("Schnelles") und etliche Diskussionen über Gott und die Welt. Ebenso seien hier Christian Spengler, Carlos Tellaeche, Philipp Günther, Andreas Freund, Holger Aschenbrenner, Vanessa Gepert, Jens Bernnat und Winfried Säckel erwähnt, die entscheidend zur positiven Arbeitsatmosphäre am ICVT und damit zum Entstehen der Arbeit beigetragen haben.

Dr.-Ing. Gheorghe Sorescu als Administrator und Institutsfinanzfachmann danke ich unter anderem für die Unterstützung bei administrativen Angelegenheiten. Nicht unerwähnt darf Katrin Hungerbühler als große Hilfe beim Verwalten und Abrechnen von Projekten bleiben.

Die vorliegende Arbeit wurde hauptsächlich durch die deutsche Forschungsgemeinschaft über das internationale Graduiertenkolleg NUPUS finanziert. Besonderer Dank gilt hier Prof. Dr.-Ing. Rainer Helmig, der mit seiner offenen, freundlichen und sehr motivierenden Art den Austausch innerhalb des Kollegs immerzu gefördert hat. Maria Costa Jornet als gute Seele des Kollegs danke ich für die liebenswerte Beantwortung und Hilfe bei offenen Fragen. Andreas Lauser als "Mit-NUPUS'ler" sei herzlich für den Darcy-Code und den Support in DuMu^X-Fragen gedankt.

Zum Gelingen der Arbeit haben auch studentische Arbeiten und Hilfskräfte beigetragen: exemplarisch möchte ich hier Jeremias Bickel, Manuel Hirschler und Lishu Xiang nennen, welche mich eine längere Zeit begleitet haben.

Bei Nithart Grützmacher möchte ich mich besonders bedanken – die geleisteten Anmerkungen und Korrekturen als fachfremder, aber dennoch sehr interessierter und kritischer Leser, waren immer sehr hilfreich.

Meinen Eltern danke ich für die fortwährende Unterstützung und die Ermöglichung meines Studiums. Besonderer Dank gilt auch Jana, Christoph, Annette, Ralf und Jörg – als Freunde haben sie für viele schöne Stunden abseits des Instituts gesorgt und damit auch die notwendige Ablenkung geschaffen.

Tiefster Dank gilt meiner Frau Judith, die mich immerzu loyal unterstützt und auf viele gemeinsame Stunden verzichtet hat. Ihre Ruhe und ihr Zutrauen haben motivierend über den einen oder anderen zähen Punkt hinweggeholfen und die Fertigstellung der Arbeit ermöglicht.

Reutlingen, im Januar 2016   Stefan Dwenger

Contents

List of Figures

List of Tables

Abbreviations and List of Symbols

Abbreviations

μ-CT	Micro Computer Tomography
BPP	Bipolar Plate
CPU	Central Processing Unit
GDL	Gas Diffusion Layer
MEA	Membrane Electrode Assembly
MM	Mathematical Morphology
MPI	Message Passing Interface
NMR	Nuclear Magnetic Resonance
PEMFC	Polymer Electrolyte Membrane Fuel Cell
PTFE	Polytetrafluoroethylene
REV	Representative Elementary Volume
SA	Simulated Annealing
SEM	Scanning Electron Microscopy

Latin Letters

A	area	m^2	$[L^2]$
A	prefactor	$1/mol \cdot s$	$[1/NT]$
c	concentration	mol/m^3	$[N/L^3]$
c	heat capacity	$J/kg \cdot K$	$[L^2/T^2\Phi]$

D	(effective) diffusion coefficient	-	[-]
D	diffusion coefficient	m^2/s	$[L^2/T]$
d	distance	m	[L]
d	fibre diameter	m	[L]
d	thickness	m	[L]
E	energy	J	$[ML^2/T^2]$
e	structuring element	-	[-]
G	interfacial free energy	J	$[ML^2/T^2]$
h	specific enthalpy	J/kg	$[L^2/T^2]$
i	phase	-	[-]
J	molar flux	$mol/(m^2 \cdot s)$	$[N/L^2T]$
K	intrinsic permeability	m^2	$[L^2]$
k	(phase) permeability	m^2	$[L^2]$
k	(relative) permeability	-	[-]
k	Boltzmann constant	J/K	$[ML^2/T^2\Phi]$
l	characteristic length	m	[L]
m	mass	kg	[M]
m	van Genuchten parameter	-	[-]
n	number of CPUs	-	[-]
n	van Genuchten parameter	-	[-]
p	pressure	Pa	$[ML/T^2]$
Q	image	-	[-]
q	source term	$kg/(m^3 \cdot s)$	$[M/L^3T]$
R	universal gas constant	$J/mol \cdot K$	$[ML^2/T^2N\Phi]$
r	fibre radius	m	[L]
r	radius	m	[L]
S	saturation	-	[-]
T	temperature	K	$[\Phi]$
t	time	s	[T]
u	internal energy	J	$[ML^2/T^2]$
V	volume	m^3	$[L^3]$
v	velocity	m/s	[M/T]
x	length	m	[L]

Greek Letters

α	contact angle	°	[-]
α	van Genuchten parameter	-	[-]
χ	voxel size	m	[L]
δ	dilation	-	[-]
ϵ	erosion	-	[-]
ϵ	porosity	-	[-]
ϵ	relative error	%	[-]
η	dynamic viscosity	Pa · s	[M/LT]
Γ	configuration of elements	-	[-]
γ	interfacial free energy	J/m^2	$[M/T^2]$
λ	heat conductivity	W/m · K	$[ML/T^3\Phi]$
λ	mean free path	m	[L]
λ	mobility	$1/(Pa · s)$	[LT/M]
λ	pore-size distribution index	-	[-]
μ	chemical potential	J/mol	$[ML^2/T^2N]$
ν	kinematic viscosity	m^2/s	L^2/T
Φ	volume	m^3	$[L^3]$
ϕ	porosity	-	[-]
ρ	density	kg/m^3	$[M/L^3]$
σ	standard deviation	-	[-]
σ	surface tension	N / m	$[M/T^2]$

Superscripts

nw	non-wetting
s	system
w	wetting

Subscripts

a	activation
a	area
ana	analytical
b	bubbling
c	capillary
crit	critical
d	entry
dis	discretized
eff	effective
fac	factor
gas	gaseous
i	phase
ini	initial
j	component
liq	liquid
new	new
P	PTFE
porous	porous medium
r	relative
r	residual
S	solid
t	total
V	vapour
W/w	water

Dimensionless numbers

Ca	Capillary number	$Ca = \frac{\eta \cdot v}{\sigma}$
Kn	Knudsen number	$Kn = \frac{d}{\lambda}$
Re	Reynold number	$Re = \frac{\rho \cdot v \cdot d}{\eta}$

Zusammenfassung

In der vorliegenden Arbeit wird der Einfluss der wässrigen Flüssigphase und Gasphase auf das Transportverhalten in Gasdiffusionsmedien der PEM-Brennstoffzelle simulativ und experimentell untersucht. Zu Beginn werden die Grundlagen von Polymerelektrolyt-membran–Brennstoffzellen, die für künftige mobile und stationäre Anwendungen eine mögliche saubere Energiewandlung darstellt, erläutert. Darüber hinaus werden Vor- und Nachteile von PEM-Brennstoffzellen dargestellt und diskutiert. Alle Bestandteile der Brennstoffzelle (Membran mit Katalysator, Gasdiffusionsmedien, Bipolarplatte mit durchströmter Struktur) und das im Betrieb auftretende und für die Leistung entscheidende Wassermanagement werden beschrieben. In den letzten Jahren wurde immer deutlicher, dass die Transportprozesse von Wasser und Gas in den gemischt benetzbaren Gasdiffusionsmedien eine entscheidende Rolle für das Verständnis dieser Art von Brennstoffzellen spielen. Ein kurzer Überblick über bestehende Techniken zur Darstellung des Gehalts und der Verteilung des Wassers wird ebenfalls gegeben. Zusätzlich werden die verschiedenen Transportmechanismen und deren Einfluss auf die Wasserverteilung in der Zelle, sowie die Abbildung in theoretischen Modellen besprochen. Im Besonderen werden die so genannten REV-basierenden Darcy-Modelle und deren Grundlagen, beginnend von den verschiedenen Größenskalen (Micro- vs. Makroskala), über die dadurch entstehenden gemittelten Größen auf der REV-Skala (representative elementary volume) hin zu zusätzlichen Annahmen und Limitierungen hervorgehoben. Ebenfalls werden die entstehenden Konstitutivbeziehungen und deren Abbildung in verschiedenen Modellen (generell, als auch spezifisch für GDLs) beschrieben. Abschließend werden die Grundgleichungen für solche REV-Modelle, die später zur Beschreibung der Transportprozesse verwendet werden, abgeleitet.

Im Anschluss wird die Modellierung und Simulation der Wasserverteilung in gemischt benetzbaren porösen Medien beleuchtet. Im vorliegenden Fall wurde ein Ansatz verwendet, der auf thermodynamischen Größen basiert — die Wechselwirkung zwischen der Gas-, Flüssig-, Feststoff- und PTFE-Phase wird mittels eines Schemas abgebildet, welches die Grenzflächenenergien zwischen den Phasen betrachtet und minimiert. Dafür muss die poröse Struktur der Gasdiffusionslage (GDL) rekonstruiert und diskretisiert werden (Algorithmen zur Generierung der Faseranordnung, die Verteilung der PTFE-Phase darin und die entstehenden Fehler bei verschiedenen Auflösungen werden ebenfalls dargestellt). Darüber hinaus ist ein passendes Optimierungsschema zur Lösung eines solchen globalen Minimierungsproblems notwendig; hierfür wurde der Simulated-Annealing-Ansatz gewählt und implementiert. Der Ansatz wurde dann mit unterschiedlichen Testfällen validiert: zu Beginn wurden verschiedene Randbedingungen getestet, anschließend wurde ein Beispiel mit Röhren erzeugt und abschließend konnte mittels zwei sich schneidender Fasern (mit und ohne hydrophober Beschichtung) gezeigt werden, dass der Ansatz die auftretenden Phänomene in gemischt benetzbaren porösen Schichten abbilden kann.

Am Ende des Kapitels wird noch eine parallele Ausführung des Simulated Annealing dargestellt. Hierzu wird zu Beginn ein Überblick über bestehende Parallelisierungstechniken für Simulated Annealing Algorithmen mit Fokus auf problemunabhängigen Verfahren gegeben. Zusätzlich wird die Zerlegung der Modellierungsdomain für die Energieoptimierung innerhalb der Gasdiffusionslagen erläutert und in den Algorithmus, welcher am Ende des Abschnitts beschrieben wird, integriert.

Basierend auf den Ergebnissen aus dem vorhergehenden Kapitel wird die Modellierung von Konstitutivbeziehungen und Transportparametern, abhängig von PTFE-Gehalt und Phasenanteil der Flüssigphase (Sättigung), durchgeführt und beschrieben. Dafür wurde eine Methodik entwickelt, die aus folgenden Teilschritten besteht: Strukturgenerierung, Simulation der Wasserverteilung, Erkennung von Grenzflächen, Überführung in ein lesbares Format zur Gittererzeugung, Berechnung der gesuchten Größen mit openFOAM und abschließender Analyse der Simulationsdaten mittels ParaView. Ergänzend wurden Experimente zur Messung dieser Transportbeziehungen entwickelt. Dabei lag der Fokus auf der definierten Kompression der betrachteten Probe und deren Auswirkung auf die Kapillardruck-Sättigungsbeziehung (p_c-S_w), auf Permeabilitäten, relative Permeabilitäts-Sättigungsbeziehung (k_r-S_w) und den effektiven Diffusivitäten (D_{eff}). Hierzu wurden Messzellen entwickelt, welche die Möglichkeit eröffnen, die Kompression mikrometergenau einzustellen, was dann die Imitierung der Kompressionsbedingungen in Brennstoffzellen erlaubt. Zudem wurden entsprechende Betriebs- und Konditionierungsstrategien entwickelt, um damit ein wiederholbares Messprinzip zu ermöglichen. Im Fall der Permeabilitäten und Diffusivitäten wurde der anisotrope Charakter der Gasdiffusionsmedien in Betracht gezogen. Hierzu wurden verschiedene Apparaturen für die Messungen quer und längs zur Hauptrichtung der GDL-Fasern dargestellt. Stationäre Methoden zur Bestimmung der relativen Permeabilitäten (Penn-State-Methode) und effektiven Diffusivitäten (Wicke-Kallenbach-Zelle) wurden angewandt. Basierend auf den experimentellen und simulierten Ergebnissen konnte eine verlässliche Basis zur Simulation des Gegenstromprozesses in gemischt benetzbaren Medien dargestellt werden.

Mit Hilfe der ermittelten Konstitutivbeziehungen und Transportparametern wurde ein REV-basierendes Darcy-Modell erstellt und mit den oben beschriebenen Transportgrößen versehen. Ein Vergleich gegenüber einem integralen Experiment, welches durch die Strömungsbedingungen in der Kathode der Brennstoffzelle (Gegenstromprozess von Gas- und Flüssigphase in gemischt-benetzbaren GDLs) motiviert ist, erfolgte parallel. Das Experiment wurde hierfür in mehreren Schritten entwickelt — das finale Design erlaubt die Darstellung eines Gegenstromtransportprozesses von Gas und flüssigen Wasser in porösen Medien in Kombination mit einer katalytischen Reaktion, welche Wasser unter definierten Bedingungen (isotherm bei jeglichen Temperaturen, definierte Feuchte) erzeugt. Mit Hilfe einer Lanze konnten axiale Konzentrationsprofile im durchströmten Kanal gemessen werden; zusätzlich war eine optische Beobachtung der GDL von oben möglich.

Die Ergebnisse der Reaktorexperimente wurden zur Prüfung, ob das 2D-Darcy-Modell das Verhalten des Reaktors wiedergeben kann, herangezogen. Hierzu wurde das 2D-Modell mit den abgeleiteten Konstitutivbeziehungen und Transportparametern aus dem zweiten und dritten Kapitel ausgerüstet. Die darin verwendeten Modelle werden normalerweise bei großskaligen Fragestellungen (Geophysik, hier Kilometerskala) angewandt, wohingegen die Prozesse in der GDL auf einer sehr feinen Skala (Mikrometer) auftreten. Im vorliegenden Fall wurden die gemessenen Konzentrationsprofile bei verschiedenen Bedingungen durch die durchgeführten Simulationen bestätigt. Ergänzend erlauben diese einen ersten detaillierten Blick auf den Wassergehalt bzw. die Wasserverteilung innerhalb der gemischt benetzbaren porösen Medien bei definierten Gegenstrombedingungen von Gas- und Flüssigphase.

Abstract

In this thesis the influence of liquid water and gas on the transport properties of gas diffusion media of polymer electrolyte membrane fuel cells (PEMFC) is examined numerically and experimentally.

First, fundamentals of PEMFC as one promising clean power source for mobile and stationary applications are presented. Moreover, advantages and disadvantages of PEM fuel cells are given and discussed. All the parts of the fuel cell (membrane with catalytic active substances, gas diffusion medium, bipolar plate with flow field) and the arising challenging water management are sketched out and described. In the last decades it has become clear that the transport of gas and water inside the mixed-wettable gas diffusion medium plays a significant role for the improved understanding of this type of fuel cells. A short overview of experimental techniques for the determination of the water content and the distribution inside the cell is also given. Additionally, the different transport mechanisms and their influence on the water profile across the cell and their representation in theoretical models are presented.

Especially the so-called REV-based Darcy models and their fundamentals, starting from different spatial scales (micro vs. macro scale), and the averaging of quantities leading to the representative elementary volume (REV), up to additional assumptions and limitations, are highlighted. The arising constitutive relationships and transport parameters linked with hysteretic behaviour and their representation in several models (in general as well as for GDLs) are also described. Finally, fundamental equations for such REV models, which, be later, are used for the description of transport processes, are derived and given.

Subsequently, the modelling and simulation of the water distribution inside mixed-wettable porous media, especially in gas diffusion layers, is discussed. To this end, a thermodynamical-based approach is chosen — the interactions between gaseous, liquid, solid, and PTFE phases are treated with the help of a stationary scheme based on the interfacial energies which have to be minimized. For that the porous structure must be generated and discretized (algorithms for the generation, the distribution of water-repellent PTFE coating, and evolving errors for different resolutions are discussed). Moreover, an appropriate optimization scheme is needed to solve such a global minimization problem. Here, the simulated annealing approach is chosen and implemented. The approach is then validated in different test cases: initially different boundary conditions are tested, then a synthetic example representing tubular pores with different wettability is generated, and finally it is pointed out that two intersecting fibres with and without coating showed that the approach captures the arising physical phenomena of mixed-wettability inside porous gas diffusion layers. At the end of the chapter the parallel version of a simulated annealing scheme is also discussed. Different existing parallelization techniques are reviewed and the class of problem-independent approaches is characterized in more detail. Additionally, the description of the domain decomposition for the energetic optimization inside virtual gas diffusion media is given and incorporated into the final algorithm, which is presented at the end of that section.

Based on the results of the previous section the modelling of constitutive relationships and transport parameters depending on PTFE content and water saturation inside the structure is performed. For this purpose a special method is developed and established: the structure generation, then the simulation of the water distribution, then the detection of interfacial areas and the transition to readable file formats for the meshing process, then the computing of the quantities of interest with the help of OpenFOAM, and, finally the analysis of the simulation data with ParaView.
Moreover, experiments for the measurement of these relationships are developed. A special focus is on the compression of the sample and the resulting change of the capillary pressure–saturation relationships (p_c-S_w), permeabilities, relative permeability–saturation relationships (k_r-S_w), and the effective diffusivities (D_{eff}). Therefore special measurement cells are developed including the possibility of moving in micron steps to mimic the real fuel cell conditions. Appropriate pre-conditioning and operation strate-

gies are also developed to enable a repeatable measurement procedure. In the case of permeabilities and diffusivities the anisotropic character of gas diffusion media is taken into consideration: different apparatus for in- and through-plane measurements are developed. Steady state methods for the determination of relative permeabilities (Penn State method) and effective diffusivities (Wicke-Kallenbach cell) are applied. Based on the experimental and numerical results a reliable basis for the simulation of counter-current flow processes in mixed-wettable porous media is given.

With the help of the derived transport parameters and constitutive relationships a REV-based Darcy model is set up and compared with an integral experiment, which is motivated by the flow regime at the cathode side of a polymer electrolyte membrane fuel cell where the gas and liquid phase are flowing inside the mixed-wettable gas diffusion media in a counter-current process.

The experiment is developed in several steps. The final design enables the establishment of a counter-current flow process of gas and water inside mixed-wettable porous media combined with a catalytic-induced reaction forming water under controlled conditions (isothermal at each temperature, defined water humidity). With the help of a lance, axial gas concentration profiles inside the flown-through channel can be measured. Moreover, a visual observation of the gas diffusion layer from the top of the reactor is possible.

The experimental results are taken to prove whether the developed 2D REV-based Darcy model, which is set up and later on equipped with derived constitutive relationships and transport parameters from the second and third chapter, is able to reproduce the behaviour of the reactor. Such models are normally applied to large-scale problems (geophysics, scales of km), whereas the processes in the GDL occur on a fine scale (microns). In the present case the measured concentration profiles under different conditions are confirmed with the conducted simulations. They also enable a first detailed view of the water content and its distribution inside thin mixed-wettable porous media under defined counter-current flow regimes of the gaseous and liquid phase.

1
Motivation and Introduction

In times of growing world population and increasing demand of usable energy as well as limited reserves of fossil energy, alternative forms of energy supply have to be developed. Fuel cells as a promising power source for several fields of application like mobile devices, new car concepts, auxiliary power units for military purposes or civil use (e.g. in yachts or motor caravans) as well as decentralized power and heat generation are not well established yet due to high costs, limited lifespan and high demands on peripheral devices. Nevertheless, as fuel cell techniques have promised an attractive future for different applications in the last two decades, modelling and experimental research have been increased substantially.

In the first section of this chapter fundamentals of polymer electrolyte membrane fuel cells and their challenging water management as a basis for further considerations will be explained. The second section describes the basics of porous media as well as the framework for modelling transport processes in porous media. For the modelling of counter-current flow of gas and water in the gas diffusion layer as a mixed-wettable porous medium a Darcy-flow based approach is chosen. Moreover, fundamentals, different scales, requirements, assumptions, and existing approaches of constitutive relationships as characteristics for transport properties of gas diffusion media are sketched out and elucidated. The corresponding equations for conservation of mass, momentum, and energy complete the given framework. Additionally, hysteretic behaviour of capillary pressure is addressed. A brief summary of the complete thesis is placed at the end of the chapter.

1.1 Fuel Cell Fundamentals

The interaction of electro-chemical processes, thermodynamics and transport processes on different scales in a fuel cell makes the detailed description of the complete system a real challenge. A huge effort has been taken to investigate and describe these processes in fuel cells in great detail. On the one hand a large number of experimental investigations have been conducted, on the other hand different modelling concepts have been applied to fuel cells.

The polymer electrolyte membrane fuel cell (PEMFC) as one exponent of the different types of fuel cells available will be characterized in the following more in detail. *Larminie and Dicks* [59] give a good overview about the principles of fuel cells in general as well as the design and application of the respective systems.
W. T. Grubb and L. Niedrach, working for GE in the early sixties, developed the first fuel cell with an ion exchange membrane utilized later in the Gemini programme of the NASA. Nowadays, the polymer electrolyte membrane fuel cell enables us to run the hydrogen-oxygen reaction producing pure water, electricity and heat under controlled conditions (cf. also reaction equation below).

$$
\begin{array}{lll}
\text{Anode:} & 2\,H_2 & \rightarrow \quad 4\,H^+ + 4\,e^- \\
\text{Cathode:} & O_2 + 4\,H^+ + 4\,e^- & \rightarrow \quad 2\,H_2O \\
\hline
\text{overall reaction:} & 2\,H_2 + O_2 & \rightarrow \quad 2\,H_2O
\end{array}
$$

The cell itself consists of different types of components: the membrane, catalytic active substances (platinum or a mixture of platinum and ruthenium on a carbon support), the diffusion media, and the flow field for the supply of gases as well as the removal of produced water. Figure 1.3 shows a schematic sketch of a single cell as a cut-out of a stack. The combination of the membrane and the catalytic layer (depicted as light-grey) in the middle is embraced by gas diffusion layers (dark-grey). The feed gases oxygen and hydrogen as well as the cooling (often a mixture of glycol and water) are transported to the cell via the bipolar plates neighbouring the gas diffusion media. The flowing media are fed through the manifolds (rectangular holes at the sides of the plates in the sketch) which are surrounded by a sealing (black line) similar to the active area in the middle of the cell.

At elevated temperatures the feed gases are pre-humidified to prevent the cell, especially the membrane, from drying out. All the components of a single cell as well as stack-specific parts will be described in the following more precisely. The peripheral devices, such as humidifier, compressors, pumps or ejectors etc. will not be addressed; for details refer to *Larminie and Dicks* [59] or *O'Hayre et al.* [74] for example. The

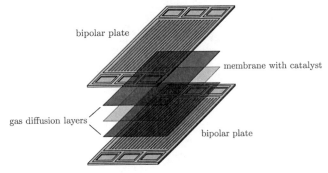

bipolar plate

membrane with catalyst

gas diffusion layers

bipolar plate

Figure 1.1: *Principle sketch of a single fuel cell*

combination of multiple cells in a stack provides high electrical currents while the compact design is preserved. Subsequently, each component of a single cell or a stack is described in detail.

Surveying a stack, the outermost components which can be seen are the end plates. They have to ensure an even pressure distribution all over the stack as well as a possibility of introducing the flowing media (gases and cooling) into the manifolds of the bipolar plates. The end plates also incorporate the tie anchors and any length compensation elements which are responsible for the force application to the stack. Hence, a stiff and lightweight design is aspired. Actually the end plates are made up of steel or aluminium, which are typically galvanically passivated to ensure compatibility to gases and cooling. Moreover the electrical contact resistance is minimized. The serial alignment of single cells results in an electrical positive and negative contact located at the edge cells of the stack. Therefore the end plates normally connect the stack to the electrical consumer.

Another problem is the sealing of the stack. Multiple bipolar plates including manifolds for the media and large areas adjacent to the atmosphere necessitate sealing concepts which are life-time free of leakage. Moreover oxygen, hydrogen, and the cooling fluid inside the cells have to be separated strictly to avoid an undesired harmful oxyhydrogen reaction or accumulation inside the cooling circuit. The sealing compound used has to be chemically compatible to the gases as well as to the cooling fluid. Moreover, thermal and mechanical loads have to be tolerated. The sealing can be applied directly both to the membrane electrode assembly and to the sub-gasket (which is mostly a thin film) or to the bipolar plates. Other solutions like inserting O-rings or sealing elements in precast grooves are not applicable to large-scale production.

The bipolar plate (BPP) is necessary to connect the single cells to each other in a stack in serial mode. These stacks often consist of up to 400 cells to increase the output voltage (especially in the automotive industry). On the one side of the bipolar plate the cathode of a single cell is attached and on the other side the anode of the adjacent cell. Besides the bipolar plate separates the two different gas supplies and ensures the proper provision of reactants. The total flux of reactant gases is fed through channels at the side of the stack embedded in each bipolar plate forming the previously mentioned manifolds. From the manifolds hydrogen and oxygen are fed through coined or milled channels (depending on the material of the bipolar plate). The entity of channels is known as flow field. Several different general designs of flow fields are known. Figure 1.2 shows the different possibilities of designing them.
The parallel pattern (fig. 1.2(a)) is the simplest, but has the great disadvantage that water, which might be blocking several channels, is not forced to migrate. Generally speaking, the shifting of a droplet sticking in a channel is determined by the balance of forces: convective forces, i.e. pressure drop of flowing gases, and surface tension compete against each other. If the convective force is bigger than the surface tension, the droplet moves and, consequently, is removed from the flow field. If the water is not removed, the cell output decreases due to non-supplied areas where the gas flow is hindered by the stagnant water. This critical point is solved by serpentine flow fields (fig. 1.2(b)), where a single channel is flown through: occurring water blockages are removed by the gas stream. The enormous length of a single channel leads to unmeant high pressure drops. The combination of both types is the parallel serpentine flow

field (fig. 1.2(c)): a continuous supply of gases is secured and the resulting length of the channels is admissible. In contrast to the types specified above of flow fields the interdigitated pattern (fig. 1.2(d)) leads to an enforced flow of gases through the gas diffusion medium. This is due to the dead-end of each channel: the inflowing gases are forced to intrude into the gas diffusion medium as a porous structure, flowing through it under the ribs of the bipolar plate and leaving it into the outflow channels of the flow field. Especially on the cathode side of the polymer electrolyte membrane fuel cell the reaction product water is removed from the GDL and carried out with the help of this type of flow field and thus the flooding of the cell is prevented and an efficient supply of reactants to the catalytic layer is ensured. Unfortunately, as a result of forced convection the pressure drop over the cell is increased strongly.

During the selection process for an adequate flow field the width of the channels and ribs (covering a portion of the GDL) has to be taken into account with respect to volume fluxes. The channels on the cathode side are often deeper and expanded in comparison to the anode side. The manufacturability of such fine details and their function in support of the compression of the cell and the resulting pressure drop have also to be regarded.

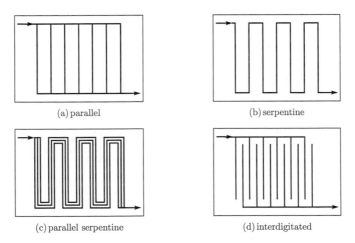

(a) parallel   (b) serpentine

(c) parallel serpentine   (d) interdigitated

Figure 1.2: *Different types of flow field designs for PEMFC*

To ensure sufficient electrical conductivity the assembly of electrodes and membrane is slightly pressed between the bipolar plates with their flow fields. For stacks a thermostatization is often required: so that the interior of each plate can be flown through with a coolant helping to maintain a desired temperature level. The plates are usually manufactured from graphite or galvanically (e.g. with gold) passivated stainless steel. The chosen material has to be chemically inert to media flowing through the bipolar plate. The optimal plate is very thin with the result that the construction volume is small and the entire stack of little weight. It also has low electrical resistivity and at the same time only small points of connection to the electrodes to facilitate high provision of reactants, which leads to discrepancies. These aspects have to be solved during the design and construction phase of a bipolar plate.

The membrane electrode assembly (MEA), which consists of the membrane, the catalyst, and the gas diffusion layer, is the central part of the polymer electrolyte membrane fuel cell. These individual parts of the MEA are generally hot-pressed to ensure a deep connection. The partial reactions themselves (cf. also page 2) are running at the three-phase boundary where the membrane, the catalyst, and the reactant gas are present. This is why the three-phase boundary should have a maximal surface. In the following each part of the MEA will be described in detail.

The membrane in the middle of the two sides of a fuel cell has several tasks - on the one hand it ensures that the two different gases are strictly separated. Separation requires a gas-tight membrane. On the other hand the membrane has to be proton-conductive but impermeable for electrons. In addition to that a high chemical and thermal stability is required because of the operation conditions present in fuel cells. For polymer electrolyte membrane fuel cells the membrane has a matrix consisting of a polymer which is modified by hydrophilic sulphonic acid groups. Therefore the polymer matrix is normally treated with sulphuric acid. The resulting functional groups enable the dissociation of protons by SO_3-ions while water is present. As a consequence the protons are able to migrate then from the anode to the cathode side of a PEMFC (cf. also figure 1.3). Migration of protons is in first-order proportional to the water content of the membrane [65]. The water-uptake of polymer membranes can reach a value of 20 % of weight. The most popular material for polymer electrolyte membranes is Nafion, a perfluorsulphonic acid polymer developed in the sixties.

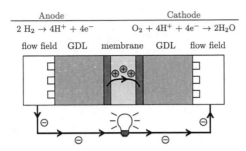

Figure 1.3: *Electro-chemical reactions and charge flow of a PEMFC*

The occurring chemical reactions are catalysed by noble metals which are dispersed in small particles (diameter in the range of some nanometres). Nowadays platinum is a standard material. The combination with ruthenium as a blend is also possible and increases the tolerance of the cell against carbon monoxide. To achieve a maximal specific surface, the catalytic active particles are divided over a carbon support. A mixture of those carbon particles, PTFE particles, and polymer membrane is processed in a further step. Including particles of the membrane into the catalytic active layer often leads to an improved proton conductivity of the catalytic layer. Two ways are generally known for the deposition: the separate electrode method and building the electrode directly onto the electrolyte [59]. With the separate electrode method the carbon support (and if necessary PTFE particles to enhance water repellency which leads to increased rates of water removal from the catalyst) is transferred for example by spraying it onto the surface of the gas diffusion medium. The diffusion medium with the active particles is joined with the membrane and hot-pressed for a certain time resulting in the so-called MEA. An alternative route is the fixing of the particles (noble metals, carbon support, hydrophobic PTFE, and membrane) directly onto the membrane: rolling, printing, and spraying techniques are well-known.

The gas diffusion layers, the material of foremost interest for the present thesis, located between the gas flow field and the membrane, are normally made of textile mats composed of carbon fibres. Two main types of gas diffusion layers can be distinguished: non-wovens (also known as carbon paper, cf. figure 1.4(a), 1.4(b)) and woven (carbon cloth, cf. figure 1.4(c), 1.4(d)) gas diffusion media. Both are highly anisotropic, i.e. the fibres are mainly orientated within the x-y-plane which, later on, is parallel to the catalytic layer within the assembled fuel cell. The gas diffusion layer has to connect

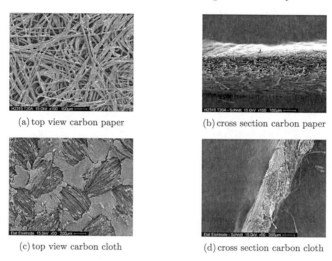

(a) top view carbon paper   (b) cross section carbon paper

(c) top view carbon cloth   (d) cross section carbon cloth

Figure 1.4: *Different types of gas diffusion media*

electrically the carbon-supported catalyst and the bipolar plate with each other. Also, the GDL conveys the heat produced by the electro-chemical reaction at the catalyst to the bipolar plate. The mechanical protection of the sensitive membrane and the catalyst is a further task of the diffusion layer. Normally the gas diffusion medium of the anode side is slightly thicker than the medium at the cathode side to prevent the membrane from drying out. The fibrous assemble is hydrophobized with e.g. PTFE (a white covering layer of the fibres, cf. figure 1.4 or characterizations of GDLs in the appendix on page 113) to create a water-repellent surface which enhances the water removal from the catalyst as well as the diffusion of the reactants. The hydrophobic PTFE particles are suspended in water and transferred to the GDL via spraying or dipping. A drying process followed by sintering at elevated temperatures (approx. 300 °C) fixes the PTFE on the surface of the fibres. Several groups have reported that the GDL is not strictly hydrophobic [31, 38]: some areas are not covered with the hydrophobic coating linked to the coating process. As a consequence a mixed wetting behaviour appears: parts of the GDL are strictly hydrophobic, others are hydrophilic.

In the following some advantages as well as drawbacks of polymer electrolyte membrane fuel cell systems will be sketched out. Amongst other things the PEMFC offers high efficiency, also under dynamic conditions, in comparison to internal combustion engines for example. Additionally, the PEM fuel cell itself owns few moving parts and has a solid membrane as electrolyte so that it can prevent leaks of harmful substances. Subsequently, the operation of fuel cell stacks do rarely emit noise. The cathode of the cell itself can be run with pure oxygen as feed gas instead of air, which results in high current densities ($> 1\,^{A}/_{cm^2}$). The catalytic induced hydrogen-oxygen reaction produces only harmless water as a reaction product - no emissions to the environment will occur. These facts emphasize the potential for long-lasting and reliable usage of PEMFC systems.

On the other hand some disadvantages have to be mentioned: the polymer electrolyte membrane fuel cell needs highly purified educts. In case of impurity (e.g. traces of carbon monoxide, ammonia or sulphur compounds) the membrane or the catalytic converter are damaged or polluted. Especially the production of pure hydrogen is at the moment expensive and complex. Today hydrogen is gained as a side product in refineries or directly by the reforming of hydro-carbons (also producing carbon dioxide as an atmospheric greenhouse gas). In the future hydrogen might also be produced by electrolysis powered by solar electricity. Furthermore, at the moment, the components included in the PEMFC stack are comparatively expensive - but, on the other hand increasing sales output will potentially beat down the prices of these relative complex systems in the future.

The most important point for polymer electrolyte membrane fuel cells is the complicated water management in the cell, i.e. the membrane as an ionomer (combination of non-polar thermoplast with polar monomers, for example DuPont's Nafion) is only proton-conductive if a certain amount of water is present (roughly the proton conductivity is proportional to the water content of the membrane). By contrast, liquid water in the adjacent gas diffusion layer blocks gaseous reactants migrating to the catalytic layer. For an optimal performance of the PEM fuel cell a balance of sufficient water in the membrane and little water in the diffusion media is needed. Depending on feed gas concentration and humidification level, the reaction rates in the cell are locally varying, resulting in different amounts of liquid water inside the cell. Normally, the stack is driven at elevated temperatures to increase reaction rates and reduce losses such as cathode activation or (over-)voltage drops. Therefore the stack temperature is generally in the field of 80 °C, but to eliminate drying effects at elevated temperatures (above 60 °C the incoming air at the cathode dries out the electrodes), a humidification of feed gases is generally needed. Though, on the other hand, this results in an extra source of water within the system. Hence, the optimal humidification level is about 80 % rH, otherwise liquid water accumulates and can block the gas diffusion media or channels of the flow field. The pinning of water droplets in the channels can be minimized by an appropriate design of the flow field (cf. figure 1.2 on page 5) or by establishing a pressure drop of the flowing gases between inlet and outlet of the flow field which is greater than the surface tension of the water droplets.

The water management of the PEMFC is mainly influenced by the feed gas properties as shown above, but also strongly by transport phenomena in the cell itself. What are the transport processes? Figure 1.5 shows a cross-section through a single fuel cell including the bipolar plate with the flow field, the gas diffusion layer, and the membrane in the middle. Moreover a profile of the water content is indicated (black line). As depicted in the cross-section of a single PEM fuel cell several major transport

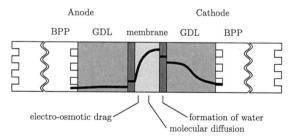

Figure 1.5: *Cross-section through a single cell with a profile of water content*

mechanisms can be observed: diffusion, capillary forces, and the electro-osmotic drag. Inside the membrane water diffuses from one electrode to the other so that a flux to the anode as well as a flux to the cathode side of the cell can be observed. The direction depends on the water content at each side of the membrane – the flux is defined by the addiction of the water to balance the water content across the membrane. Capillary forces, which will be addressed in more detail later on, are also present. The two transport phenomena mentioned are based on concentration and saturation gradients of the species. Moreover pressure-induced convective water transport is present but underpart. However, the electro-osmotic drag describing the transport of water molecules from the anode side to the cathode of a fuel cell caused by migrating protons plays an important role. *Zawodzinski et al.* [105] report that up to five water molecules per moving proton are dragged from the anode to the cathode of the fuel cell. At high current densities this phenomenon might lead to a dry anode side of the polymer electrolyte membrane.

The water forming reaction takes place at the three-phase boundary (location where membrane, catalyst, and reactant gas are present) of the cathode. As a consequence the water content of the cathode in general is almost always higher compared to the anode side of the PEM fuel cell. Therefore, the gas diffusion media are filled with water. If the vapour or liquid water phase is primarily present, depends on the load of the cell – as a rule of thumb you can say that high loads lead to more liquid water, especially in the gas diffusion layer of the cathode.

Due to the strong influence of the water content on the power output and performance of the polymer electrolyte membrane fuel cell, the water management accordingly looms large. Modelling these processes might be one approach to improve the understanding of the whole system - quite a number of authors have setup models trying to describe the complete fuel cell (i.e. mass transport in the membrane and the electrodes as well as the electrochemistry), for example *Meier* [65], *Ochs* [73], *Wieser* [99], *Wöhr* [103]. With regard to these models, several open questions arise concerning the detailed physics in the gas diffusion layer on the one hand and the coupling of the porous media with the gas channels on the other hand. Often empirical approaches are chosen to overcome the lack of physical models, which leads to phenomenological or strongly simplified relationships included in the models. Consequently these approaches cannot predict the system behaviour, e.g. in case of alternation of load, in general. The aforementioned authors included those phenomena for instance with sorption approaches fitting the measured datasets of testing systems in the lab. The resulting models capture the system performance of the considered setup quite well, but changes of materials, operation conditions etc. cannot be predicted correctly.

In order to describe and forecast processes and electro-chemistry in a PEM fuel cell the mass transport of gases and water as the key process must be grasped in more detail. Especially the counter-current transport of gas and (liquid) water in the gas diffusion layer has to be regarded exactly and resolved locally to broaden the understanding. Experimental observation and investigation of phenomena occurring on the pore scale level is rather difficult due to the tiny dimensions and the relatively fast processes (e.g. the wetting of pores is done in the range of seconds). Highly sophisticated techniques like micro computer tomography (μ-CT) or nuclear magnetic resonance techniques (NMR) are expensive, difficult to handle and sometimes insufficient to capture all the processes of interest. Therefore, more precisely, rather different modelling approaches have been implemented and presented in literature to enlarge the knowledge of processes like counter-current flow in the gas diffusion media of polymer electrolyte membrane fuel cells. Concepts such as pore network models [40], where the porous medium is described as an ensemble of throats and pores, M^2-models [96] including viscous coupling of two phases by additional cross-terms or models using Darcy flow REV-based concepts [3, 73] have been set up in the last 20 years for example. The latter class of models will be addressed in the following in more detail.

1.2 Fluid Transport in Porous Media

Acosta et al. [3] have introduced the class of REV-based Darcy models to fuel cell applications, where REV stands for the representative elementary volume. In contrast to pore network models where the structure of the GDL as well as transport processes are locally resolved by adequate geometrical approximations and simplifications, an averaging of microscale phenomena leads to macroscopic quantities. In the subsequent paragraph, assumptions, fundamentals, terms, and definitions as well as equations for REV-based models incorporating Darcy-flow will be given and discussed. Points of contact to other applications in general and especially in process engineering will also be presented. In general *Bear* [11] as well as *Bear and Bachmat* [12] give a comprehensive overview of fundamentals, definitions and concepts of flow in porous media. The intention of this paragraph is to explain the chosen approaches, definitions, and spatial levels in order to provide a common basis for the following equations and paragraphs.

Porous media consist of a solid matrix and a system of pores. Depending on the chosen point of view different physical phenomena can be observed. On the molecular level the interactions of molecules with each other are regarded: the insert in figure 1.6 depicts some water molecules interacting with a solid surface. Nowadays, due to the enormous number of molecules (1 mol incorporates 6.02×10^{23} particles), simulations of flow and transport in porous media on the molecular level are not possible. The same applies to the position and direction of movement for each molecule as initial conditions for simulations are not observable. Thus, the required set of information is not completely available. Despite these limitations the molecular scale plays an important role: properties of gases, liquids, and mixtures such as viscosity, density or diffusion mechanisms and coefficients as well as wettability phenomena can be explained with the help of considerations on the molecular scale.

The pore scale with microscopic quantities is based on a continuum approach where the ensemble of molecules provided by the molecular level is averaged. Consequently, the microscale approach arises where the fluid and solid phase are regarded as continuum including physical and dynamical properties of the phase so that the molecular point of view vanishes and direct measurable quantities arise. The amplitude of the curve given in figure 1.6 representing the fluctuations of each phase over space is consequently getting smaller. On the microscale the phases can be clearly separated (under the assumption that the areas of interest are small enough); the motion of fluids through pores on the microscale can be specified by the NAVIER-STOKES equation for example.

At this stage the flow and transport through porous media can be described with the help of fluid mechanics, but the relevant computations are not feasible due to the enormous amount of information needed for a complete description of occurring processes in irregular porous media. These deficiencies and drawbacks can be compensated by a new averaging process - the macroscale with the representative elementary volume arises. As a consequence the actual porous medium is replaced by an artificial

continuum owning new properties [11]: discontinuities observable on the microscale
will disappear on the REV-scale, which is often referred to as the macroscale. The
averaging process is only applicable to extensive quantities like mass, volume etc. During the averaging, characteristic properties on the microscale have to be preserved and
assigned or transferred to the macroscale. Figure 1.6 shows the schematic plot of an

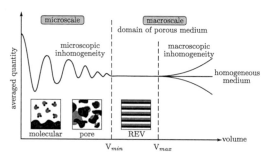

Figure 1.6: *Definition of REV and scales [11]*

averaged quantity over the averaging volume. At the very beginning of the curve a
fixed value is detectable - for very small averaging volumes, going also along with the
molecular point of view, the fluctuations are very large. Enlarging the averaging volume and incorporating the continuum approach, the pore scale as a microscale arises.
Different phases i (gaseous, solid, liquid) with phase-dependent properties are clearly
distinguishable from each other at this stage. Crossing V_{min} (cf. axis of ordinates in
figure 1.6) the averaged quantity takes a constant value. The macroscale or the domain of porous medium begins at this threshold. V_{min} represents the lower boundary
of an appropriate representative elementary volume – expanding the averaging volume
leads to the upper boundary value V_{max} for the REV. Between V_{min} and V_{max} averaged
quantities are sufficiently independent of the size of the considered averaging volume.
The averaging volumes are often referred to as characteristic lengths. In the case of
heterogeneous porous media a further growing averaging volume applied to porous
media shows a deviation from the straight, horizontal line in the middle representing
a homogeneous medium.

For REV-models phases are defined as homogeneous entities with different properties
building sharp interfaces between each other. Moreover, every phase can consist of

several components. For the description of the phases and their interactions among each other in the system, the assumption of thermodynamical equilibrium is chosen. Three pillars together form the thermodynamical equilibrium, which will be shortly sketched out in the following for a two-phase multi-component system ($i = 1, 2$ and j species) [92]:

- thermal equilibrium: $T^1 = T^2$

- mechanical equilibrium: $p^1 = p^2$

- chemical equilibrium: $\mu_j^1 = \mu_j^2$

Due to the relatively slow processes in porous media the criteria are fulfilled on the REV-scale. The mechanical equilibrium is also valid in porous media in case of consideration of capillary pressure, expressed as a pressure jump at the occurring interface of the two phases. In the present case two phases occur: the liquid phase, denoted as *liq*, and the gaseous phase, labelled *gas*.

The REV-based Darcy-flow approach is well known in soil mechanics or reservoir engineering, where the systems of soils are almost all hydrophilic [11, 12, 26, 46, 85]. In the present case, however, the considered gas diffusion media as a technical porous media has a (partly) hydrophobic character. Independent of the type of porous medium (homogeneous vs. heterogeneous matrix or wettability), the consecutively described framework of assumptions and equations is also applicable to drying processes, filter papers and fibrous aggregates as well as flow and transport in bulk good for example. A fundamental point is the description of the considered porous medium: only macroscopic, i.e. averaged quantities are suitable for Darcy-flow based modelling. Resolving phenomena on the pore-scale with the help of these model concepts is, by definition, not possible (cf. also figure 1.6).

The Darcy-flow approach itself is characterized by the incorporation of surface tension and a pore radius dependent capillary pressure as a driving force for transport processes. Especially for fuel cells, the capillary-driven transport processes are of substantial interest. The capillary forces and interactions can be characterized with the help of the dimensionless capillary number Ca expressing the ratio of viscous forces related to surface tension. With respect to the capillary number Ca, the flow in the gas diffusion media of PEM fuel cells is mainly determined by capillary forces (i.e. Ca number is below 1×10^{-6}) because the flow velocity of the water phase inside the gas diffusion layers is small.

$$Ca = \frac{\eta \cdot v}{\sigma} \tag{1.1}$$

The dynamic viscosity is given by η, the velocity is denoted with v, and the surface tension with σ as an acting force at the interface between the liquid (water) phase and gaseous phase.

One crucial relationship for transport models of porous media by the generalized Darcy-approach is the capillary pressure-saturation relationship. The filling and draining behaviour of liquid water inside the porous material can be described by p_c-S_w curves. When talking about capillary pressure some differentiations should be made depending on the point of consideration. Starting with the microscale, the classical definition of the capillary pressure depending on the curvature is given by:

$$p_c^{micro} = \sigma \cdot \left(\frac{1}{r_x} + \frac{1}{r_y} \right) \tag{1.2}$$

where r_x and r_y are the main radii of curvature. In case of a circular duct the LAPLACE-equation results:

$$p_c^{micro} = \frac{4 \cdot \sigma \cdot cos\ \alpha}{d} \tag{1.3}$$

Here d denotes the pore diameter, σ is the surface tension and α the contact angle. The existing two different phases (gas and liquid water) can be distinguished locally in contrast to the macro scale model where averaged quantities are used (cf. for example [46]). The gap between the micro scale and the macro scale is bridged by an averaging process. The averaging process and the underlying REV are sketched on page 14. As a result of the averaging process locally resolved information like pore shape and surface properties disappears but new quantities arise: saturation S_i of phase i and macroscopic capillary pressure p_c. The macroscopic capillary pressure is defined as the pressure difference between the non-wetting phase p^{nw} and the wetting phase p^w:

$$p_c = p^{nw} - p^w \tag{1.4}$$

In the present thesis liquid water is defined as the non-wetting phase nw whereas the gaseous phase is assumed as the wetting species. The definition of wettability is done to emphasize the hydrophobic character of gas diffusion media.

The saturation S_i of the phase i is generally defined as:

$$S_i = \frac{\Phi_i}{\Phi} = \frac{\text{volume of fluid phase i in domain}}{\text{volume of pore space in domain}} \tag{1.5}$$

As mentioned above, the capillary pressure-saturation relationship (p_c-S_w) gives information about pressures for the filling and draining of the porous body depending on the amount of liquid water inside the pores. Unfortunately these relationships are not biunique - different characteristics arise for wetting and drainage. The so-called hysteresis is one crucial point for p_c-S_w-curves. Figure 1.7 shows a sketch of a capillary pressure-saturation relationship including imbition and drainage curves as well as scanning curves between them.

Several reasons for hysteresis are known and described in the literature [12, 46]: contact angle, ink-bottle effect, and residual saturation. Contact angle hysteresis (sometimes also described as raindrop effect) is due to the fact that advancing and receding contact angles of fluids can differ from each other, additionally the roughness and properties of the surfaces play also an important role for the wetting behaviour. Different positions of the decisive radii inside the pore system occur for the imbition or drainage process leading to the so-called ink-bottle effect. Moreover, the residual saturation S_{ir} describing the irreducible amount of phase i trapped inside the porous body is given in the sketch. The matter for residual saturation can be found during drainage - certain portions of liquid are restrained and kept immobile inside the porous medium. The residual saturation of water inside gas diffusion layers tends to be zero, which will be described in more detail later on. As a result of the described phenomena the capillary pressure during imbition is not equal to the capillary pressure while draining the porous system. The incorporation of hysteresis between imbition (wetting) and

drainage (drying) is a central issue, because capillary pressure is the main driving force
for the considered processes in the gas diffusion layer of the fuel cell. The influence

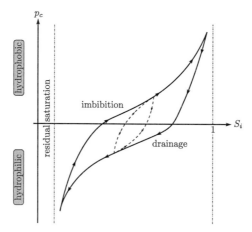

Figure 1.7: *Hysteresis of capillary pressure*

of different stages of compression and of the content of water-repellent PTFE inside
the gas diffusion media onto the shape of the curves must also be examined. The
relevant investigations will be presented in chapter 3, including the description of the
experimental setup and the presentation of results.

Several researchers have developed parametrizable models for the convenient description of p_c-S_w-curves. The most popular approaches in geoscience for non-hysteretic behaviour are the LEVERETT [63], BROOKS AND COREY [16] and the VAN GENUCHTEN
[37] models. Hysteretic behaviour of p_c-S_w-curves has been addressed by *Parker et al.*
[82] - in opposition to play-type models where only the main imbibition and drainage
curves (solid lines in fig. 1.7) are used, Parker-Lenhard models capture the physics
more accurate due to the absence of interpolation functions for scanning curves in the
area between the main curves. Instead of interpolation functions the main imbibition
and drainage curve are scaled each time according to the effective saturation and a
current residual saturation inside the porous medium. With this approach the capillary pressure for the desired saturation can be evaluated. Details about the application

and implementation of the Parker-Lenhard model can be found, for example, in *Lauser* [60]. In the following, non-hysteretic models for two-phase systems will be presented.

- **J-Leverett function**:

$$J(S_w) = \frac{p_c \cdot \sqrt{\frac{K}{\phi}}}{\sigma} \qquad (1.6)$$

- **Brooks and Corey**:

$$S_{eff}(p_c) = \frac{S_w - S_{wr}}{1 - S_{wr}} = \frac{p_d{}^\lambda}{p_c} \qquad \text{if } p_c \geq p_d \qquad (1.7)$$

$$p_c(S_w) = p_d \cdot S_{eff}^{-\frac{1}{\lambda}} \qquad \text{if } p_c \geq p_d \qquad (1.8)$$

- **van Genuchten**:

$$S_{eff}(p_c) = \frac{S_w - S_{wr}}{1 - S_{wr}} = [1 + (\alpha \cdot p_c)^n]^m \qquad \text{if } p_c > 0 \qquad (1.9)$$

$$p_c(S_w) = \frac{1}{\alpha} \cdot \left[S_{eff}^{-\frac{1}{m}} - 1 \right]^{\frac{1}{n}} \qquad \text{if } p_c > 0 \qquad (1.10)$$

K represents the permeability of the porous medium and ϕ the porosity. The effective saturation S_{eff} is calculated via the residual saturation S_{wr} of the wetting phase w. The J-Leverett functions are scaled, semi-empirical functions (based on dimensional analysis) where the interactions between the fluids and the matrix are enclosed via K, ϕ, and σ. Several researcher have reported that the contact angle between the fluids and the matrix additionally plays an important role, so they added a function depending on the contact angle to the J-Leverett function. The BROOKS&COREY parameterization uses λ as a parameter (which describes the pore size distribution: small values represent a uni-modal distribution, big values indicate a broad distribution) as well as the entry pressure p_d, which corresponds to the largest intruded pore in the porous ensemble. The parameters α, n, and m are taken to form the VAN GENUCHTEN equation. Normally these aforementioned equations are fitted to data from measurements of capillary pressure-saturation relationships.

Helmig [46] as well as *Bear and Bachmat* [12] give good overviews of these models, which mainly refer only to hydrophilic porous media, whereas the gas diffusion layer of a polymer electrolyte membrane fuel cell is mixed wettable, i.e. parts of the porous body own hydrophilic properties in contrast to hydrophobic portions. As a result these parametrizations are not able to capture the physics of a mixed-wettable porous medium like the gas diffusion medium of a PEMFC.

Nam and Kaviany [70] as well as *Pasaogullari and Wang* [83] have therefore introduced a J-Leverett function for mixed-wettable gas diffusion layers. Depending on the contact angle, the parametrization is modified:

$$J(S_w) = \begin{cases} 1.417 \cdot (1 - S_w) - 2.120 \cdot (1 - S_w)^2 + 1.263 \cdot (1 - S_w)^3 & \text{if } \alpha < 90° \\ 1.417 \cdot S_w - 2.120 \cdot S_w^2 + 1.263 \cdot S_w^3 & \text{if } \alpha > 90° \end{cases} \quad (1.11)$$

The adapted J-Leverett function is taken to calculate the capillary pressure depending on porosity, permeability, and contact angle of the gas diffusion medium:

$$p_c = \sigma \cdot \cos \alpha \cdot \sqrt{\frac{\phi}{K}} \cdot J(S_w) \quad (1.12)$$

The approach is easy to adapt to existing systems or modelling tasks but fails in case of materials with other structural properties. New types of GDLs and different applied compression loads cannot be described accurately with such a general approach. Moreover, the phenomenon of hysteresis is not covered. Therefore measurements and new approaches are required. In chapter 3 techniques for the precise description of gas diffusion media are presented.

As shown with equation (1.1) on page 16, the processes in the backing of the fuel cell are primarily driven by capillary processes. However, convective processes, which dominate for example under the ribs of the bipolar plate in case of an interdigitated flow field, must also be regarded. The interaction of the gaseous and liquid phases while flowing in porous media is one crucial point for large-scale as well as technical applications like the electrodes of PEM fuel cells. The macroscopic description of flow in and through porous media is mainly done with the (extended) Darcy law and material balances. In the following paragraph fundamentals of Darcy's law and necessary parameters like the relative permeability and belonging parametrizations will be discussed.

Darcy's law is only valid for pore Reynold numbers smaller than approximately one [26], which is also known as creeping flow.

$$Re = \frac{\rho \cdot v \cdot d}{\eta} \tag{1.13}$$

Creeping flow is additionally limited with respect to phenomena on the molecular level. This is why the Knudsen number has been defined and in general a lower limit of 10 is chosen (limit for continuum mechanics). Here, λ represents the mean free path of gas molecules inside a pore with diameter d.

$$Kn = \frac{\lambda}{d} \tag{1.14}$$

If the restrictions are fulfilled Darcy's law can be applied. The single-phase Darcy law is given by:

$$v = \frac{K}{\eta} \cdot \frac{\Delta p}{d} \tag{1.15}$$

where K is the permeability tensor, also known as DARCY permeability. The permeability tensor is defined as:

$$\mathbf{K} = \begin{bmatrix} k_{xx} & k_{xy} & k_{xz} \\ k_{xy} & k_{yy} & k_{yz} \\ k_{xz} & k_{yz} & k_{zz} \end{bmatrix} \tag{1.16}$$

If a Cartesian coordinate system is chosen for the description of flow in an anisotropic porous medium, the tensor K is symmetrical. For isotropic porous media the three elements of the principal diagonal are given by the scalar k.

The extended Darcy law for two phase flow under steady-state conditions with respect to phase i reads as follows:

$$v_i = \frac{k_i}{\eta_i} \cdot \frac{\Delta p_i}{d} \tag{1.17}$$

v_i is the velocity of phase i, k_i represents the effective permeability, which is also known as phase permeability. Pressure drop Δp, thickness d of the flown-through porous material, and dynamic viscosity η complete the equation. The Darcy equations for the different phases are linked with each other via the capillary pressure p_c (cf. equation (1.4)).

Relative permeability k_{ri} of phase i is defined as the ratio of phase permeability k_i and intrinsic permeability K:

$$k_{ri} = \frac{k_i}{K} \tag{1.18}$$

The value of the relative permeability has no dimension and depends only on the saturation of the porous medium:

$$0 \leq \sum_{i=1}^{n_{phase}} k_{ri}(S_i) \leq 1 \tag{1.19}$$

Intrinsic permeability K, in contrast, is only up to the material. Normally the determination of K is based on the measurement of pressure drop Δp at a precisely defined volume flux $\dot{V}$ under knowledge of flown-through cross-section A and thickness d of the porous sample. With the help of the following formula the intrinsic permeability can be calculated:

$$K = \frac{\dot{V} \cdot \eta \cdot d}{\Delta p \cdot A} \tag{1.20}$$

In contrast, the shape of the relative permeability-saturation curve (cf. also figure 1.8) depends on several properties: the structure of the porous medium as well as the interactions between the two phases with each other (which also strongly depend on the saturation level) and with the pore structure have an influence on the relative permeability itself. Combining equations (1.17) and (1.18) yields to the reformulated extended Darcy law, where the fraction $\frac{k_{ri}}{\eta_i}$ is also known as mobility λ_i.

$$v_i = \frac{k_{ri} \cdot K}{\eta_i} \cdot \frac{\Delta p_i}{d} \tag{1.21}$$

Again, several approaches for the parametrization of relative permeability-saturation curves on the macroscale are given in the literature. Subsequently a couple of those

will be presented and discussed. *Bear* [11] as well as *Scheidegger* [85] give a broad
summary of basis research on relative permeability, especially for two-phase systems,
which will be addressed here. Relative permeability k_{ri} of phase i acts as a saturation-
depending scaling factor in the multiphase Darcy law and represents the interactions
of different phases on each other. With respect to the underlying assumptions, an
exchange of momentum between the phases is negligible [12].

Similar to capillary pressure-saturation relationships, hysteresis can occur – different
shapes of curves are observable depending on the process (imbibition vs. drainage).
In terms of gas diffusion media and the interaction of water and gas as a two-phase
system, the hysteretic behaviour tends to be small [46]. Generally four concepts for

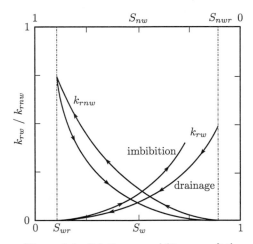

Figure 1.8: *Relative permeability curves [11]*

the description of relative permeability-saturation curves can be distinguished: capillaric, statistical, network models, and empirical relations [11, 26]. For further details
concerning capillaric and network models refer to *Dullien* [26]. The class of statistical
models (especially pore network models) is for example represented by *Burdine* [19],
who has developed expressions for relative permeabilities based on the bundle of capillary tubes model where the porous medium is fictional given as ensemble of capillaries.
Assuming Hagen-Poiseuille flow inside the tubes and averaging the velocities over the

cross-section in combination with Darcy's law ("hydraulic radius theory"), the terms
for relative permeabilities can be derived:

$$k_{rw} = S_{eff}^2 \cdot \frac{\int_0^{S_{eff}} \frac{dS_{eff}}{p_c^2(S_{eff})}}{\int_0^1 \frac{dS_{eff}}{p_c^2(S_{eff})}} \tag{1.22}$$

The relative permeability for non-wetting phase nw is derived analogously:

$$k_{rnw} = (1 - S_{eff})^2 \cdot \frac{\int_0^{S_{eff}} \frac{dS_{eff}}{p_c^2(S_{eff})}}{\int_0^1 \frac{dS_{eff}}{p_c^2(S_{eff})}} \tag{1.23}$$

Empirical relations are almost always motivated by experimental results - in the present
case *Brooks and Corey* [16] have shown the validity of their approach for different types
of porous isotropic media. The equations for wetting phase w and non-wetting phase
nw are given:

$$k_{rw} = S_{eff}^{\frac{2+3\cdot\lambda}{\lambda}} = \left(\frac{p_b}{p_c}\right)^{(2+3\cdot\lambda)} \tag{1.24}$$

$$k_{rnw} = (1 - S_{eff})^2 \cdot (1 - S_{eff}^{\frac{2+\lambda}{\lambda}}) = \left[1 - \left(\frac{p_b}{p_c}\right)^\lambda\right]^2 \cdot \left[1 - \left(\frac{p_b}{p_c}\right)^{(2+\lambda)}\right] \tag{1.25}$$

Bubbling pressure p_b as well as pore-size distribution index λ are determined with the
help of experiments.

As discussed in the previous paragraph, the Darcy law represents the equation for
the momentum. For modelling purposes a reasonable mathematical representation of
the processes in the porous medium has to be defined. Therefore basic equations for
each phase / component will be defined and appropriate initial and boundary conditions will be set. The explanations in the following will be restricted to two-phase
systems. The following momentum balances, continuity equations as well as energy
balances are restricted to systems with thermodynamical equilibrium (cf. page 15).
Moreover, creeping flow must be present ($Re < 1$): the definitions (1.13) and (1.14)
have to be fulfilled. Finally, by assuming that intertia effects are negligible, external

forces are constant, the solid phase is immobile, and the fluid phases are macroscopically frictionless [46], then the subsequent framework of equations is able to represent flow and transport in porous media systems.

Applying the Eulerian point of view where the control volume (REV) does not change in time and space, the continuity equation with a storage term, a convective part, and a source or sink term is given and holds the criteria of mass conservation. For each phase a continuity equation can be formulated reading as follows:

$$\frac{\partial(S_i \cdot \phi \cdot \rho_i)}{\partial t} + \nabla \cdot (\rho_i \cdot v_i) - \rho_i \cdot q_i = 0 \tag{1.26}$$

The conservation of momentum as the second quantity of classical mechanics is introduced via the extended Darcy law (cf. equations (1.15), (1.17)). The combination of the extended Darcy law (1.21) and the continuity equation (1.26) leads to the general differential equation for multiphase flow in porous media. Additionally, the following assumptions (resulting from the first term of (1.26)) are also applied to the continuity equation:

$$\rho_i \cdot S_i \cdot \frac{\partial \phi}{\partial t} = 0 \qquad \text{i.e. incompressible matrix} \tag{1.27}$$

$$\phi \cdot S_i \cdot \frac{\partial \rho_i}{\partial t} = \phi \cdot S_i \cdot \frac{\partial \rho_i}{\partial p} \cdot \frac{\partial p}{\partial t} = 0 \qquad \text{i.e. incompressible fluids} \tag{1.28}$$

The highly non-linear differential equation for multiphase flow in porous media is given in the following:

$$\phi \cdot \rho_i \cdot \frac{\partial S_i}{\partial t} - \nabla \cdot \left[k_{ri} \cdot \frac{\rho_i}{\eta_i} \cdot K \cdot (\nabla p_i - \rho_i \cdot g) \right] - \rho_i \cdot q_i = 0 \tag{1.29}$$

With the constraints of

$$\sum_{i=1}^{n_{phase}} S_i = 1 \tag{1.30}$$

and the capillary pressure

$$p_c(S_i) = p^{nw} - p^w \tag{1.31}$$

Moreover, for non-isothermal systems the conservation of energy is required. In equation (1.32) the specific enthalpy of phase i is balanced. For a detailed derivation of the equation see *Helmig* [46]. Heat capacity c in the first term of the equation and heat conductivity λ in the last are averaged quantities that depend on the saturation and the porosity of the matrix. Specific enthalpy h_i is given by the sum of internal energy u_i and $\frac{p_i}{\rho_i}$. Via q_i as thermal source or sink term, incorporation of thermal energy into the framework is realized. Dissipation between the phases is due to creeping flow negligible; additionally thermal equilibrium between the solid matrix and the fluids is assumed [46].

$$
\begin{aligned}
0 = \sum_{i=1}^{n_{phase}} \int_G \Bigg[& (1 - \phi) \cdot \rho_s \cdot c \cdot \frac{\partial T}{\partial t} + \phi \cdot \frac{\partial \left(u_i \cdot \rho_i \cdot S_i \right)}{\partial t} \\
& - \nabla \cdot \left(v_i \cdot \rho_i \cdot \left(u_i + \frac{p_i}{\rho_i} \right) \right) \\
& - \nabla \cdot \left(\lambda \cdot \nabla T - \sum_{k=1}^{k_{comp}} \rho \cdot g \cdot D \cdot h_g^k \cdot \nabla X_g^k \right) - q_i \Bigg]
\end{aligned}
\tag{1.32}
$$

The presented system of equations is highly coupled and non-linear due to the constitutive relationships for the relative permeability and the capillary pressure, which both depend on the saturation. Adequate discretization schemes and solutions strategies will be presented in chapter 4.

The aforementioned model concept relies on averaged or effective material parameters and constitutive relationships like (relative) permeabilities depending on saturation, capillary pressure-saturation relationships, pore size distribution, wettability, shape and composition of the material etc. Several of the said parameters and relationships are provided by literature or manufacturers. Others like the compression-depending p_c-S_w or k_r-S_w relationships are little published and known in the literature. Therefore adequate and precise measurement and modelling strategies have to be developed and validated.

Spatial variant porous media systems (e.g. different types of media or cracks and fractures) are challenging tasks for the averaging process - upscaling and averaging for them and their constitutive relationships and parameters are a matter of actual research [72]. Therefore the question arises if the gas diffusion layers as subjects of interest in the present case can be regarded as a porous medium which is applicable to models depending on averaged quantities. Moreover, the validity and application of the Darcy-based equations (resting upon the described averaging process) for the given counter-current transport process have to be proven.

1.3 Outline of the Thesis

The application of the framework of equations given in the first chapter highlights the overall goal of this dissertation - the possibility to model and simulate the transport of multiphase multi-component mixtures in GDLs as porous media without adjustable parameters. In the case of REV-based models using a generalized Darcy-approach, which are regarded here in detail, constitutive relationships and parameters have to be determined experimentally and with the help of simulations. The capillary pressure-saturation relationship as well as the relative permeability-saturation relationships including the intrinsic permeability, are crucial relationships for the transport models of the gas diffusion layer. Assumptions, terms, and required definitions for these models are also illustrated in this chapter. In the present case, the knowledge is applied to the processes in the gas diffusion layer of a polymer electrolyte membrane fuel cell, whose fundamentals are given here too. Figure 1.9 on page 28 illustrates the complete workflow of the present thesis.

In the second chapter the modelling of the water distribution in mixed-wettable porous fibre assembles is described (depicted also in the right area in figure 1.9). It acts as a basis for further numerical determinations of transport parameters and constitutive relationships. The approach as well as the chosen optimization schedule are given and discussed. Elementary investigations have been done concerning the shape and structure of fibrous porous materials with respect to the resulting water distribution.

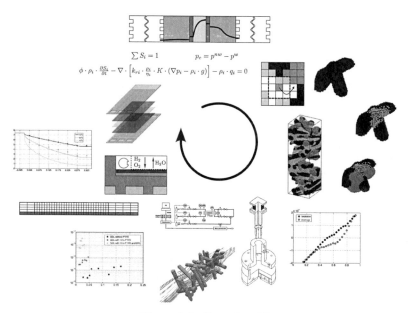

Figure 1.9: *General workflow*

In chapter three resulting water distributions in virtually generated gas diffusion media are taken to calculate transport parameters like relative permeability or effective diffusivity. Experimental set-ups for the determination of capillary pressure-saturation relationships, relative permeabilities, and effective diffusivities are presented (cf. also figure 1.9). Measured results are checked against the numerical determined data-sets.

In chapter four the equations for the framework of equations for REV-based Darcy models given in the first chapter as well as the results from the third are implemented and taken to investigate counter-current flow of multi-component gas mixtures and liquid water in thin hydrophobic porous media (i.e. in the present case commercial gas diffusion layers). Additionally an experiment with counter-current flow of gas and water in thin porous media was installed and operated. Then, again, a comparison of experiments with simulations is performed to validate the chosen approach.

The conclusions at the end of the thesis sum up the results and make possible suggestions on the future work. In the appendix gas diffusion media are characterized in detail and algorithms for the virtual material design of GDLs, which provide a basis for the simulations, are presented. Moreover, additional results, parameters and values are given.

2

Simulation of Water Distribution

The water distribution inside the polymer electrolyte membrane fuel cell, especially inside the gas diffusion layer, has a strong influence on the transport of species inside the cell. The power output of polymer electrolyte membrane fuel cell systems strongly depends on the transport of reactants to the catalytic layer as well as the removal of the produced liquid water (cf. chapter 1). The transport of species itself inside the porous backing of the polymer electrolyte membrane fuel cell is mainly influenced by the amount of liquid water in the gas diffusion layers. Hence, the knowledge of the water distribution inside the gas diffusion layer depending on structural properties is one crucial point to improve the performance of fuel cells.

In this chapter a strategy for the prediction of the water distribution inside the gas diffusion layer will be presented, the influence of structural parameters on the resulting water content and its spreading over the fibrous mixed-wettable porous media will be shown and discussed, and afterwards the obtained structural data sets will be utilised for the simulation of the constitutive relationships and transport parameters (cf. next chapter 3).

Modelling the water distribution inside mixed-wettable porous gas diffusion layers (cf. appendix A for properties of GDLs) can be conducted in several ways. In general two classes of model approaches can be distinguished: on the one hand transient models, e.g. [97, 106], which are able to capture dynamic phenomena inside the gas diffusion layer as a porous medium as well as the transition to the gas channel, and, on the other

hand, stationary schemes such as primary pore-network models [22, 40] and stochastic
models [15, 56, 88–90, 93].

The Monte Carlo approach, described in the following, is based on thermodynamics:
Knight et al. [56] have first proposed to model the interactions between the gaseous,
liquid and solid phase as an optimization problem. The microscopic phase distribution
is simulated with the help of the interfacial free energies where the equilibrium phase
distribution matches with an energy level near or equal to the global minimum of the
interfacial free energies. Two different exemplary applications will be described in the
following: *Silverstein and Fort* [89] have utilised the resulting fluid distribution to
determine interfacial areas inside porous media; *Mohanty* [67] has applied the Monte
Carlo technique to model a fluid distribution, which subsequently was chosen to calculate the thermal conductivity of a partially saturated porous medium.

The initial step for modelling the fluid and water distribution is to discretize the area
under investigation: the virtually generated gas diffusion medium (for details of the
algorithms refer to appendix B) is regarded as an ensemble of points or particles with
specific properties (solid and PTFE). In the three-dimensional case the particles are
cubic shaped. Figure 2.1 illustrates an extract of a domain. The light-grey box represents the particle which is considered for balancing the interfacial free energies. For a

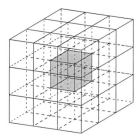

Figure 2.1: *Principle sketch of discretization*

detailed analysis of evolving discretization errors refer to section B.4, where 2D– and
3D–analyses of errors are conducted. In the following, for a more convenient illustra-
tion spheres have been chosen. Exemplary an extract of a single fibre (drawn in black)
with hydrophobic coating (light grey) is shown in figure 2.2(a).

With known geometric dimensions of the domain and discretized fibres with hydrophobic PTFE coating on the surface, the resulting void space can be calculated easily. To realize the desired water saturation S_w inside the examined domain, the number of water containing points (dark grey) has to be computed. The saturation in general is defined as given in equation (1.5). Consequentially, the water saturation for the ensemble of particles in the examined domain can be written as

$$S_w = \frac{\text{number of water particles}}{\text{number of particles in pore space}} \qquad (2.1)$$

When the number of water particles is known, they can be distributed randomly all over the void space to generate a proper starting point for modelling (cf. figure 2.2(b)). The random distribution is chosen to avoid bias or influence of the final results due to predetermined configurations. Frequently repeated optimization runs for the same geometry with different initial water distributions have verified this approach. The resulting final water distribution of the modelling process is shown in figure 2.2(c). During the modelling process the hydrophilic solid (black particles) and hydrophobic

(a) single fibre with coating   (b) initial water distribution   (c) resulting water distribution

Figure 2.2: *Example of simulation of water distribution around a single fibre*

PTFE (light-grey) particles are kept immobile while elements of liquid water and vapour change places. The produced new configuration is evaluated with respect to the interfacial energy.

Consequently, the subsequent target function for the interfacial free energies has to be minimized to find configurations with lowest energy:

$$G_t^s = \sum_i A_i \cdot \gamma_i \tag{2.2}$$

G_t^s denotes the total interfacial free energy of the global system, A_i is the surface between the discrete elements, and γ_i are the interfacial free energies for species i. If one knows the interfacial free energies for each combination of species in the considered system and minimize the proposed target function, a physically reasonable water distribution inside a porous medium can be achieved. By extending the approach to mixed-wettable porous media, an additional phase comes up. The hydrophobic part of the porous body (denoted by subscript P) has to be included in the model of *Knight et al.* [56] additionally. The hydrophilic substrate S, vapour phase V, and water phase W close the considered mixed-wettable system. For the simulation of the resulting water distribution in mixed-wettable porous systems the interfacial free energies at each interface are necessary. In the present case the interfacial free energies are given at a temperature of 25 °C, motivated by the ambient conditions given during the measurements in the laboratory (cf. chapter 3). The interfacial free energies for each combination of species are given as follows:

- **SV**: solid carbon fibres / saturated water vapour; $\gamma_{SV} = 100.36\,\mathrm{mJ/m^2}$

- **SL**: solid carbon fibres / liquid water; $\gamma_{SL} = 95.34\,\mathrm{mJ/m^2}$ [47]

- **VL**: saturated water vapour / liquid water; $\gamma_{VL} = 72\,\mathrm{mJ/m^2}$ [4]

- **PL**: solid PTFE / liquid water; $\gamma_{PL} = 50\,\mathrm{mJ/m^2}$ [50]

- **PV**: solid PTFE / saturated water vapour; $\gamma_{PV} = 22.3\,\mathrm{mJ/m^2}$ [55]

The value of γ_{SV} for the interfacial free energy of solid carbon fibres and saturated water vapour are calculated with the help of the known contact angle $\alpha_{carbon/water}$ of water-carbon systems (86°) [4] and the equation of YOUNG.

$$\gamma_{SV} = \gamma_{SL} + \gamma_{VL} \cdot cos\,\alpha \tag{2.3}$$

The contact angle $\alpha_{PTFE/water}$ for PTFE surfaces and liquid water is derived analogously with the equation (2.3). The calculated value of 112.6° for the contact angle is based on the interfacial free energies given above.

After swapping liquid and vapour elements, the produced new configuration of elements Γ_{new} is evaluated with respect to interfacial energy G_{new} and compared to the original configuration Γ_{ini} with its interfacial energy G_{ini}. Therefore the resulting interfacial energy between the considered element and its 26 neighbours is calculated. For demonstration purposes a two-dimensional example is given below in figure 2.3 where the liquid element in the dashed box is swapped with a vapour element located two sites further on the right side. As an example the interfacial energy of one element

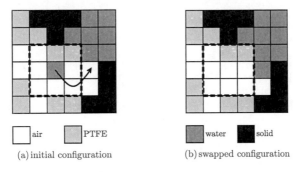

☐ air	▨ PTFE
▨ water	■ solid

(a) initial configuration   (b) swapped configuration

Figure 2.3: *Two-dimensional example for the energy balance*

with its eight neighbours inside the dashed box is considered. The initial interfacial energy G_{ini} is compared to the swapped configuration G_{new}. The difference ΔG is generally taken as a criterion of the optimization algorithm.

$$
\begin{aligned}
G_{ini} &= 2 \cdot \gamma_{PL} + 6 \cdot \gamma_{VL} &&= 2 \cdot 50\,\mathrm{mJ/m^2} + 6 \cdot 72\,\mathrm{mJ/m^2} &&= 532\,\mathrm{mJ/m^2} \\
G_{new} &= 2 \cdot \gamma_{PV} &&= 2 \cdot 22.3\,\mathrm{mJ/m^2} &&= 44.6\,\mathrm{mJ/m^2} \\
\hline
\Delta G &= G_{new} - G_{ini} &&= 44.6\,\mathrm{mJ/m^2} - 532\,\mathrm{mJ/m^2} &&= -487.4\,\mathrm{mJ/m^2}
\end{aligned}
$$

If ΔG is smaller than zero the swap is directly accepted, because the total interfacial energy G_t^s of the system decreases. To achieve the global minimum of the system moves with a positive ΔG have also to be accepted to overcome local energy minima.

Therefore a suitable optimization algorithm is needed which is able to minimize the resulting cost function given in equation (2.2). For global optimization several techniques are known, which can be classified in two branches: exact methods and heuristic search methods [30]. Exact methods, presupposing the knowledge of the solution (minimal cost function, which in the present case, is not known a priori), are, consequently, not adequate to solve the described optimization problem. And apart from that, the computational effort for exact methods increases extremely with growing numbers of unknowns. In the present case the number of unknowns is in the range of hundreds of thousands. As a consequence only heuristic search methods are applicable. Moreover the described problem is NP-hard (non-deterministic polynominal-time hard), i.e. if so, a solution cannot be achieved in polynomial time.

The class of heuristic search methods (e.g. evolutionary algorithms, tabu search, scatter search, simulated annealing, etc.) can minimize a cost function regarding the whole parameter space. However, as a negative point, they do not ensure that the global minimum can be found. Nevertheless, they almost converge at all times on a sufficient solution. These methods are often applied to combinatorial problems or systems where the number of possible solutions increases exponentially with the number of unknowns. Moreover, the simulated annealing algorithm is able to negotiate or avoid metastable configurations resulting in nonphysical solutions.

In the following section the simulated annealing algorithm as one method of optimization is considered in detail. Reasons for the choice and basics of the simulated annealing are given, followed by a brief description of the serial SA algorithm as well as a the chosen parallel version, which will be presented and discussed.

2.1 Simulated annealing

The simulated annealing (SA) algorithm which is based on the Metropolis algorithm [66] is quite a popular heuristic algorithm for optimization. SA belongs to the class of Monte Carlo Algorithms and mimics the cooling of a metal: slow annealing of the metal leads to a stable state, which is linked to minimal energy. *Kirkpatrick et al.* [54] and, later on *Cerny* [21] have applied the technique to optimization problems. Several different fields of SA applications are known today: the travelling salesman problem as a routing problem, graph theory, picture processing, very-large scale integration (VLSI) for computer chip design, applications in atomic and molecular physics, and the optimization of jobs or flight schedules for example. The main advantage of the SA algorithm is its ability to overcome larger barriers and local extrema and find the a priori unknown global minimum among several local minima. On the other hand the algorithm needs a lot of time and / or huge computing resources to find solutions with appropriate minimal cost functions.

In the following the (serial) simulated annealing algorithm are sketched out in pseudo code, test cases for the validation of the approach are given, and parallel techniques and algorithms are presented. In general, as variables are chosen the temperature T, the initial temperature T_{ini}, the critical temperature T_{crit}, the set factor T_{fac}, and the Boltzmann constant k, as well as the initial and new energies E_{ini} and E_{new}, respectively, here representing the considered total interfacial free energy G_t^s (cf. also equation (2.2)). The decisive difference in energy before and after the swap of elements is given by ΔE, which alters the overall energy of the considered optimization task. The Boltzmann constant is generally set to one, and the temperature as variable for the algorithm is dimensionless.

At the beginning of the simulated annealing, the initial energy E_{ini} of the complete system is calculated and stored. A swap leading to a new configuration of the system is performed and the resulting new ensemble is evaluated with respect to its energy. If the new configuration with its energy E_{new} is less than the initial one, the swap is accepted - otherwise a comparison with probability P given by a Boltzmann distribution (also known as Metropolis criterion [66]) decides about the acceptance of the swap.

$$P = \begin{cases} 1 & \text{if } \Delta E < 0 \\ exp\left(-\frac{\Delta E}{k \cdot T}\right) & \text{if } \Delta E \geq 0 \end{cases} \tag{2.4}$$

With increasing numbers of algorithm loops the portion of accepted energetically disadvantageous configurations (helping to overcome local maxima) decreases and so helps to find the global minimum. Thus, the temperature T must successively be lowered: therefore different schemes are available, but the most commonly used one is the shown geometric schedule where the temperature is multiplied by a constant factor, which results in an asymptotic diminution.

Listing 2.1: *Simulated Annealing algorithm*

```
1   while T > T_crit
2   {
3     calculate initial energy E_ini
4     conduct swap
5     calculate new energy E_new
6     if E_new < E_ini
7       accept swap
8     else
9       P = exp(−delta_E / (k * T))
10      if P > rand(0,1)
11        accept swap
12
13    decrease temperature T, e.g geometrical scheme: T = T * T_fac
14  }
```

In general the schedules can be classified as static cooling schedules, where the parameters cannot be changed during execution, and dynamic ones. The presented geometric schedule is the simplest exponent of static schedules. By contrast, the class of dynamic cooling schedules allows an adaptive setting of scheduling parameters, which is often accompanied with higher efficiencies of the SA algorithm, but the complexity of dynamic schedules often impede practical use. In the present case the geometric cooling schedule is chosen.

The major challenge of the simulated annealing procedure is the correct choice of the annealing parameters and the appropriate schedule type. In the following the difficulty of correct parameters will be briefly discussed. Generally it is plausible that increased initial temperatures T_{ini} lead to higher rates of acceptance and, therewith, larger increases in energy. *Aarts and Korst* [2] have reported that the choice of T_{ini} and T_{crit} is not so important, the lowering parameter T_{fac} having the main influence on the quality of the results. Generally T_{ini} (here 1500) must be high enough to enable uphill moves and T_{crit} (here 5) has to be sufficient low to freeze the system at the energetic minimum. There are no rules for the setting of the correct value for T_{fac}. The choice is strongly problem-dependent. *Silverstein and Fort* [89] chose 0.7, *Mohanty* [67] and *Berkowitz and Hansen* [15] took 0.9 for their scheduling. In the present case T_{fac} was equal to 0.9. Trial and error experiments with lower values have shown that the global energetic minimum was not reached steadily each time for several different geometries. The choice of parameters can be verified, for example, via plotting normalized energy $\overline{G_t^s}$ against saturation S_w [15, 56]. To ensure that the best solution (in case the annealing scheme leads to worse solutions), i.e. the minimal total energy G_t^s in the whole optimization process, is maintained, the overall best solution is stored with each stop. Moreover, if a certain number of consecutive steps do not improve the quality of the solution (minimal interfacial energy), it is assumed that the equilibrium is reached.

2.1.1 Validation of the approach

Validation of the approach and the ability to capture the physics of mixed-wettable porous media are the crucial points for the modelling of water distribution in gas diffusion media. Several test cases have been set up and will be presented in this paragraph. Setting up intuitive and comprehensible test cases was the main idea. First (test case A), the formation of droplets in in a hydrophobic box (light-grey particles) was tested (test case 1 in figure 2.4(a)), then the adhesion of water onto hydrophilic surfaces (black surrounding particles in 2.4(b)) was proven, thirdly, the behaviour of water on and inside mixed-wettable structures was checked (2.4(c)). In test case A each domain consisted of 20 voxels in the x- and y-dimension whereas in the z-direction a length of five voxels was chosen.

The resulting water distribution is given as a cross-section (x-y plane, z=3) and depicted in dark-grey. For each optimization the interfacial energies given on page 34 are used. The resulting configurations of the three test cases in A show that the approach is able to capture the phenomena of mixed-wettable surfaces. Perfectly round-shaped interfaces between the water and gaseous phase have not been observed in case A. The reason for this lies in the approach itself: only the 26 direct neighbours of each element are taken to calculate the difference of interfacial energy before and after a swap (cf. also figure 2.3 on page 35).

Test cases B and C, where tubes of different diameters in combination with both wettabilities are given had the dimension of 33 voxels in x- and y-direction. Set-up B was chosen to prove if smaller hydrophilic pores are filled first and tiny hydrophobic pores will be filled last. Set-up C is a further extension of B: the interaction between the wa-

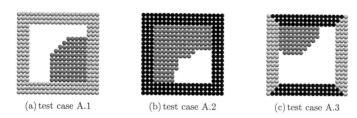

(a) test case A.1   (b) test case A.2   (c) test case A.3

Figure 2.4: *Test case A for SA validation*

ter phase (dark-grey) and the hydrophilic (black) or the hydrophobic walls (light-grey) walls of the pores was checked. To this end the hydrophilic and hydrophobic pores were merged with each other to mixed-wettable pores.

In test case B the filling of tubes is in accordance with the LAPLACE-equation given on page 16: first the hydrophilic small pores are filled with water, in the following the larger hydrophilic ones are intruded and afterwards the hydrophobic pores starting with the largest are filled. In the end the smallest hydrophobic pore is wetted.

Test case C (cf. 2.6) in comparison to set-up B shows a similar behaviour at low saturations (up to $S_w = 0.5$). If the saturation is further increased, the hydrophobic parts of the mixed-wettable pores are intruded. Similar to test case B the largest pore throat is first filled with water. Additionally ($S_w = 0.6$) the second largest hydrophobic portion of the pore is partly filled. The reason for this behaviour can be found in the energy balance at the interface of the water and solid phases. In comparison the extension of the water phase and the intrusion into the largest hydrophobic pores is more favourable as starting a new independent imbibition process into the smallest hydrophobic pore. Hence, the imbibition and drainage process and the spatial distribution of the water phase inside a mixed-wettable porous systems also depends on the spatial distribution of the two solid phases. Moreover, the adjacencies of the hydrophilic and hydrophobic solid phases have to be regarded.

Finally, test cases D.1 and D.2, which are given in figure 2.7 and 2.8, consist of two fibres. They have a diameter of 12 voxels and cross each other in a domain of the size of $30 \times 30 \times 30$ voxels. Both geometries represent an extract of a gas diffusion media where a fibrous matrix, mainly orientated in the x-y plane, builds a porous media system. Case D.1 consists of carbon fibres which are hydrophilic; D.2 shows an additional

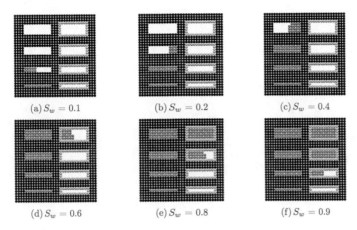

(a) $S_w = 0.1$ (b) $S_w = 0.2$ (c) $S_w = 0.4$

(d) $S_w = 0.6$ (e) $S_w = 0.8$ (f) $S_w = 0.9$

Figure 2.5: *Test case B for SA validation*

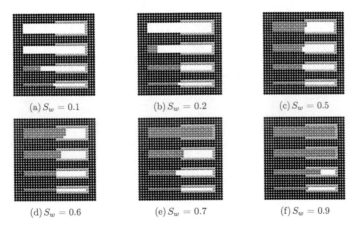

(a) $S_w = 0.1$ (b) $S_w = 0.2$ (c) $S_w = 0.5$

(d) $S_w = 0.6$ (e) $S_w = 0.7$ (f) $S_w = 0.9$

Figure 2.6: *Test case C for SA validation*

hydrophobic phase, depicted in light-grey. In the present case the hydrophobic PTFE is spread over the surface of the fibres at the intersection of both. The hydrophobic voxels are stochastically distributed (cf. also chapter B.3.2 in the appendix). In sum, 25 % of the fibre surfaces are coated with PTFE. A series of optimization runs was performed for different levels of saturation. For both scenarios the water saturation was increased in steps of 0.05. Figure 2.7 shows the empty structure as well as water

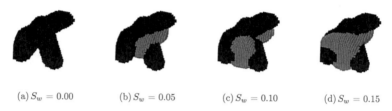

(a) $S_w = 0.00$ (b) $S_w = 0.05$ (c) $S_w = 0.10$ (d) $S_w = 0.15$

Figure 2.7: *Test case D.1: fibres without hydrophobic coating*

saturations of 0.05, 0.10, and 0.15. As a result of D.1 the water phase is mainly located at the crotch built by the intersecting fibres where the minimal total interfacial free energy level can be found. Starting with a small water film on the surface of the fibres ($S_w = 0.05$), increased water saturation leads to a significant amount of water between the two fibres.

In comparison with test case D.1 the second geometry, equipped with hydrophobic coating on the surface of the fibres, shows a different behaviour at similar water saturations (cf. also series of results in figure 2.8): due to PTFE at the intersection of the two fibres, no water in comparison to figure 2.7(b) is located there at equal water saturation levels. If water saturation is increased to $S_w = 0.15$, two separate droplets of water are formed. They will coalesce in case of further increased water saturation (figure 2.8(d), $S_w = 0.25$). Nevertheless, non-covered hydrophilic parts of the porous medium are in general preferentially wetted with water.

With the help of test cases A, B, C, D.1 and D.2 respectively, it has been demonstrated that the chosen simulated annealing approach is able to capture the physics of wettability of mixed-wettable porous media as well as their influence on the water distribution inside porous bodies. Moreover, great importance of the distribution of hydrophobic coating typically made up of PTFE for the resulting water distribution

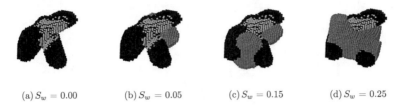

(a) $S_w = 0.00$ (b) $S_w = 0.05$ (c) $S_w = 0.15$ (d) $S_w = 0.25$

Figure 2.8: *Test case D.2: fibres with hydrophobic coating*

inside porous media can be supposed. Figures 2.9, 2.10, and 2.11 show virtually generated gas diffusion layers including hydrophobic coating at several saturations.

In the following subsection parallel techniques for speeding-up the algorithm are presented. The influence of PTFE on constitutive relationships and transport parameters is also discussed in chapter 3.

2.1.2 Parallel Simulated Annealing

The optimization of water distribution inside huge model domains with the classical simulated annealing algorithm is very time-consuming. The need of faster and more efficient solution strategies is therefore evident. The application of different scheduling schemes is discussed in the previous subsection, but the parallelization of the serial simulated annealing has not been addressed. In the following several approaches and techniques will be presented and discussed. Multiple ways of parallelization of the originally serial SA algorithm have been proposed in the literature by different authors [1, 9, 43, 62]. Some the approaches are problem-specific and will not be considered here. Nor is the differentiation between parallel approaches for shared and distributed memory architectures regarded in detail, for available actual computing platforms, e.g. clusters with hundreds of cores, lead to distributed memory capable algorithms, almost every time.

As an example, *Allwright and Carpenter* [5], in their publication have described a master-slave principle which was applied to the travelling salesman problem: worker nodes ("slaves") compute energy changes for new configurations provided by the host

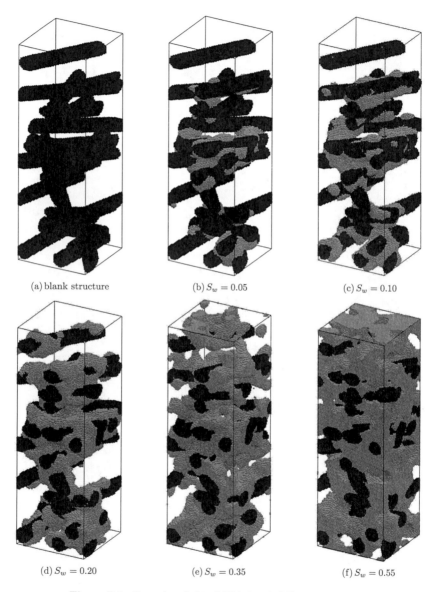

(a) blank structure (b) $S_w = 0.05$ (c) $S_w = 0.10$

(d) $S_w = 0.20$ (e) $S_w = 0.35$ (f) $S_w = 0.55$

Figure 2.9: *Examples of virtual H2315 with different saturations*

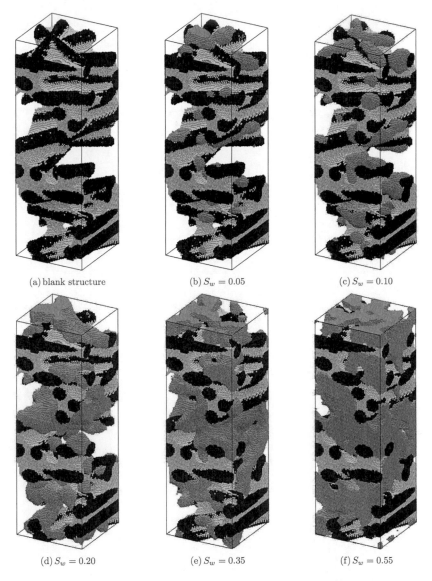

(a) blank structure   (b) $S_w = 0.05$ (c) $S_w = 0.10$

(d) $S_w = 0.20$ (e) $S_w = 0.35$ (f) $S_w = 0.55$

Figure 2.10: *Examples of virtual H2315T10A with different saturations*

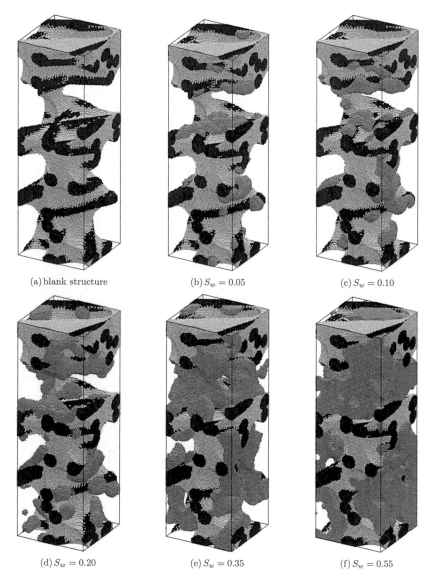

(a) blank structure   (b) $S_w = 0.05$ (c) $S_w = 0.10$

(d) $S_w = 0.20$ (e) $S_w = 0.35$ (f) $S_w = 0.55$

Figure 2.11: *Examples of virtual H2315T20A with different saturations* ("master"), and, the master node decides in a second step, if the updated constellation is accepted or discarded. Another approach was pursued by *Witte et al.* [100]: speculative computing is used to predict further steps and, consequently, speed up the algorithm. By anticipating a subset, the whole code is executed for future loops of the SA algorithm, which leads to a decrease of execution time.

Parallel execution of serial runs with and without exchange of information during the run forms another branch of parallel simulated annealing algorithms. *Aarts and Korst* [2], *Azencott* [9], and *Greening* [43] give a good overview of different possible approaches.

In general, parallelism of simulated annealing algorithms can be subdivided into two branches: single-trial and multiple-trial parallelism. Single-trial parallelism is characterized by the fact that the evolving work of the algorithm or task is divided over n available processors, which is also known as "algorithmic parallelism". In case of simulated annealing algorithms the balancing of energies, for example, can be done by a multi-core operation. Generally, it can be noticed that the potential speed-up through these operations is limited. In contrast, multiple-trial parallelism, where several trials of the SA are computed in parallel on numerous processors, offers great possibilities of computational speed-up. These approaches are often implemented in the following manner: a couple of tasks is spread over all the available CPUs and executed on each of them until the computations of one processor lead to an energetic improvement; in case of this, the solution is directly broadcasted to all the other CPUs and chosen as the initial condition for the next successional step. On the other hand this procedure results only in small speed-ups at elevated simulated annealing temperatures growing with decreasing temperatures, because the system is chaotic (initial random configuration is chosen to avoid bias) and almost every performed swap is successful and leads to time-consuming broadcast operations.

In the following, three different types of parallel simulated annealing algorithms which are not tailored to specific problems will be addressed: the systolic algorithm (sometimes also called division algorithm or independent searches), the clustering algorithm, and the error algorithm.

While using the systolic algorithm the optimization problem is divided over all the available CPUs. Two strategies are applicable. One is that each sub-problem is optimized without any communication between the processors, which results in n solutions (one for each CPU), where, in the end, the best is chosen as the final solution. The second approach owns communication between the processors during execution ("cooperating searches"). All the solutions are compared to each other and the best is broadcasted and taken as the new initial configuration during the next loop of the simulated annealing. As a result only one solution is given in the end.

With the clustering algorithm, in contrast, all the CPUs work together on the optimization problem. The problem itself can be separated into sub-problems. Working with subdivisions includes an error source of erroneous transitions due to incoherent data sets. Therefore communication between the processors of newly found solutions for the subdivisions is obligatory. In case of an accepted swap, all the other processors skip their operations, take the new configuration as initial value and generate new trials. If several accepted swaps come up in a short time, an appropriate strategy for selecting one of them has to be developed. In the present case the first incoming solution is chosen and broadcasted to all the others. Other strategies (choice of the energetically best solution, random decision or comparison with a Boltzmann distribution) were also tested, but regarding the maximum performance they were skipped.

For elevated temperatures the efficiency of the algorithm is poor due to the above mentioned reason of multiple-trial parallelism. This weakness can be eliminated with the help of an adaptive process, starting with a systolic algorithm where each processor works on its own and mutating to a classical clustering approach where all the processors exclusively work on the same data set. Figure 2.12 shows the transition between the two extrema schematically. The efficiency or the percentage of successful swaps can be chosen as a decision criterion for the mutation between the systolic and clustering algorithm.

Finally, the error algorithm is the simplest representative of parallel simulated annealing algorithms: all the available CPUs work with the same data set, which, generally, is not subdivided. No implementation of inter-processor communication of solutions is required. Moreover, accepted transitions are not broadcasted, which leads to errors: they occur for example if two processors compete against each other for the same ele-

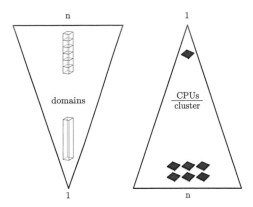

Figure 2.12: *Parallel simulated annealing domain decomposition*

ment of the data set. Consequently, the energy calculation gets wrong and, as a result, the properties of the system are changed.

Aarts and Korst [2] have tested the three different approaches with the travelling salesman problem. To this end they chose the EUR100 instance where 100 European cities with a known solution for the travelling salesman problem are represented. The result was that the error algorithm suffers under bad speed-up and growing errors while the number of processors is increased. The two other parallelisms are characterized by high efficiencies and negligible averaged errors for growing numbers of CPUs.

In the present case the systolic algorithm was first implemented and later on extended to the clustering algorithm, which, due its flexibility and adaptability, was finally applied to the optimization of the water distribution. The parallel version of the code was programmed with C using openMP shared memory parallelization (e.g. vector operations etc.) [75] and distributed memory architecture via message passing interface (MPI) (e.g. domain decomposition) [69]. The geometric cooling schedule for the parallel version of the simulated annealing is the same type as described in the previous chapter dealing with the serial algorithm.

Regarding the anisotropy of the gas diffusion layer (fibres mainly orientated in the x-y plane, cf. also the appendix on page 116ff), the spatial decomposition of the whole domain depends on the orientation of the virtual generated structure. Focusing on gas diffusion media without adjacent gas channels or ribs of the flow field, the segmentation was always performed in the z-direction. The rate of acceptance of swaps was chosen as a criterion for the rearrangement of domains during the clustering algorithm. If the acceptance rate falls below a defined value, two of initially n domains are merged into a greater domain. Successively the number of domains decreases and each domain is processed parallel by a growing number of CPUs (cf. also figure 2.12).

The applied parallel simulated annealing clustering algorithm including the adaptive domain decomposition is described as pseudo-code for n processors in listing 2.2.

Listing 2.2: *Parallel Simulated Annealing algorithm*

```
 1  do domain decomposition into n domains
 2
 3  while T > T_crit (for each task / subdomain)
 4    {
 5    calculate initial energy E_ini
 6    conduct swap
 7    calculate new energy E_new
 8    if E_new < E_ini
 9      accept swap
10    else
11      P = exp(-delta_E / (k * T))
12      if P > rand(0,1)
13        accept swap
14
15    communicate new solution to all other tasks if swap is accepted
16      skip running swaps
17      accept broadcasted new solution as initial solution
18      restart with new initial solution
19
20    if criterion < critical values && no. of domains > 1
21      resize domain
22    else
23      continue
24
25    decrease temperature T, e.g geometrical scheme: T = T * T_fac
26    }
```

To speed-up the parallel simulated annealing algorithm, the swapping of more than one pair of elements in one step is discussed, but in the present case the resulting overhead for checking if the changed pairs are independent of each other (so that no an erroneous algorithm evolves), makes the code slower in comparison to the described single swap strategies.

As depicted on pages 44 to 46 the simulation of the water inside mixed-wettable porous media is feasible. At the beginning of the chapter the approach was presented and discussed. Algorithms for simulated annealing as well as parallel simulated annealing were sketched and proven with the help of simple mixed-wettable 2D-geometries. On this basis first three-dimensional examples with two intersecting fibres without and with hydrophobic coating were set up. At the end the method was applied to virtually generated GDLs – different saturation levels were simulated and taken as a starting point for the determination of transport parameters, which are given in the following chapter. Here, the influence of structural parameters like content of PTFE or distribution of fibres onto the constitutive relationships and transport parameters will be examined and discussed. A comparison with experimentally gained values to verify the chosen approach is also done.

3

Constitutive Relationships and Transport Parameters

As sketched out in the first chapter, the description of transport processes in thin mixed-wettable porous media depends strongly on constitutive relationships and transport parameters. Darcy-flow based models (cf. chapter 1) mainly depend on the capillary pressure-saturation relationship (details presented in section 3.1), the relative permeability k_r (section 3.2), and the effective diffusivity D_{eff} (section 3.3). The main driving force for water transport in gas diffusion media of polymer electrolyte membrane fuel cells is capillarity. Moreover, the influence of the compression of the diffusion layers on the performance of the material and the behaviour inside a stack (e.g. intrusion into channels of the bipolar plate) was discussed earlier in the literature [10, 52]. It seems to be obvious that the shape of the capillary pressure–saturation curves, relative permeabilities, and effective diffusivities as constitutive relationships and transport parameters depend on the compression load. During the measurements it became finally clear that the consideration of material deformation must not be neglected. Additionally a comparison between experimental results and virtually generated materials as numerical representations is performed to identify decisive parameters and properties. Hence, the design of suitable tailored materials with increased performance capabilities may be possible in the future.

In this chapter techniques for the experimental as well as numerical determination of these relationships and parameters with and without compression will be presented and discussed. To this end, the results of the already conducted simulations of water distribution are used. In the previous chapter the methodology of the appropriate computations is presented. Figure 3.1 gives an overview of the steps of the complete process for the determination of constitutive relationships and transport parameters with the help of numerical methods. In the present case the desired material is gen-

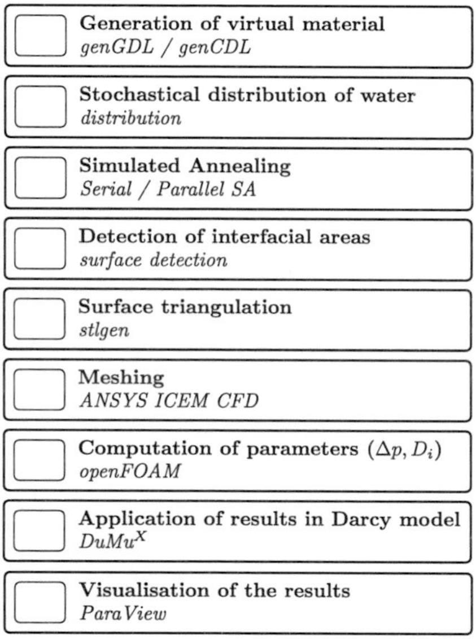

Generation of virtual material
genGDL / genCDL

Stochastical distribution of water
distribution

Simulated Annealing
Serial / Parallel SA

Detection of interfacial areas
surface detection

Surface triangulation
stlgen

Meshing
ANSYS ICEM CFD

Computation of parameters $(\Delta p, D_i)$
openFOAM

Application of results in Darcy model
DuMu^X

Visualisation of the results
ParaView

Figure 3.1: *Simulation workflow*

erated by a self-developed material generator *genGDL / genCDL* [48]. After that the chosen water saturation inside the domain is set via a random distribution function to avoid bias for the successional optimization step performed by *Serial / Parallel SA* (for details refer to chapter 2). After the optimization run (based on the simulated annealing algorithm) the interfacial areas between the water, the solid, and the gaseous phase are detected via the tool *surface detection* producing an ASCII text file with the positions of the boundary surfaces. These interfaces are the input parameters for the triangulation of the discrete material points. For this purpose, among other things, a code named *stlgen* was developed. The resulting closed surfaces are merged with a domain surrounding the virtually generated gas diffusion medium and the complete resulting stl-file is handled by *ANSYS ICEM CFD* where a 3D tetrahedral mesh is generated and exported as Fluent mesh file (*.msh). The tetrahedral mesh was chosen due to the complex geometry. The openFOAM subroutine *fluent3DMeshToFoam* converts the mesh into an applicable mesh. OpenFOAM is taken to calculate the pressure drop of the gaseous phase flowing through the (water-filled) structures and to compute the diffusion through the porous body. All calculations were performed on a standard PC. The two computed sets of information are used to determine the desired constitutive relationships and transport parameters of gas diffusion media of polymer electrolyte membrane fuel cells. As an outcome of the simulations relative permeabilities (in- and through-plane direction) for the gaseous phase as well as effective diffusivities for both main directions are available.

All the proprietary program codes described above are coded in C, the visualisation of the files is done with ParaView based on vtk-result files or special, self-defined file formats (cf. also appendix on page 126) enabling an easy to use exchange between different applications.

On the other hand the capillary pressure-saturation relationships for different gas diffusion layers with multiple levels of compression, compression-depending intrinsic permeabilities, and relative permeabilities ($k_{r,w}$) are determined by experiments. The measurement of saturation-depending effective diffusivities and the relative permeabilities for the gaseous phase ($k_{r,g}$) are not feasible due to the requirement that the water has to stay at its place during the measurements. Hence, the values are determined with the above-described simulation procedure. The details as well as the results of the simulations are presented in the next sections in parallel to the experimental setups, operation strategies, and the experimentally gained values. Later, the determined constitutive relationships and transport parameters are used for the simulation of the counter-current flow of gas and water inside thin porous layers (cf. also chapter 4).

3.1 Capillary pressure-saturation relationship

According to the general remarks in chapter one, capillary pressure is generally considered on the macro scale where an averaging process leads to REV-based quantities. The capillary pressure is almost exclusively referred to the saturation. The water saturation S_w, as a measure of the level of water inside the pores, is defined in equation (1.5). While applying Darcy-type models to fuel cell applications, the p_c-S_w function is crucial for the prediction of water and gas transport.

In the following the porous GDLs with their mixed-wettable behaviour are examined in detail. This behaviour leads to p_c-S_w diagrams where the hydrophilic character is represented by the negative capillary pressures at low saturations, whereas the hydrophobic nature of the gas diffusion layer causes positive capillary pressures. Hysteretic behaviour of capillary pressure was also measured. In 3.1.1 the experimental setup is described in detail. Then the results of several materials with different properties (blank, coated with various amounts of PTFE, finished with a micro porous layer etc.) are presented.

3.1.1 Experimental setup

The enormous influence and significance of capillary pressure-saturation relationships on the system behaviour in general, led to a large number of experiments for different types of materials conducted by generations of researchers. *Bear* [11] and *Dullien* [26] give a good overview of capillary pressure, principles of measurement, and structural effects. The present section focuses on the experimental determination of p_c-S_w relationships for gas diffusion media of PEMFC.

As a fundamental assumption in the present case, the capillary pressure is measured in steady-state conditions where the interfacial forces (expressed with the surface tension σ or the interfacial free energies γ_i, cf. page 34) are in an equilibrium state. The fundamental principle of the displacement methods was first proposed by *Bruce and Welge* [18]. They are known as "porous diaphragm method", "Welge restored state method" or as "desaturation method". Here, the fluid phases are always in hydrostatic equilibrium states. At the beginning of each measurement the sample, located between membranes to control the fluxes of different phases at the in- and outlet, is saturated with the wetting phase. Subsequently, the wetting phase is first impinged by the non-wetting phase (imbibition curve) and then vice versa (drainage curve). The capillary pressure is expressed as the pressure difference between both phases and related to the saturation of one phase, which results in p_c-S_w curves. In case of hysteretic behaviour the non-identical imbibition and drainage curves are obtained. In the present case, saturation of the water phase is denoted as S_w so that the common notation for capillary pressure–saturation relationships can be used.

In contrast to this, while using dynamic methods steady flows of two fluids (wetting and non-wetting phase) are present. *Brown* [17] built an apparatus (after Hassler's principle, which is addressed in chapter 3.2.1, in the context of relative permeability measurements) where the capillary pressure is controlled at the inlet and the outlet of the sample. Here, too, the sample is placed between two membranes which are only permeable to one fluid. The pressure difference between the two phases at the inlet is equal to the capillary pressure. If the pressure drop of both phases over the sample is identical, the sample is dismounted to determine the saturation by weighing.

Alternatively the capillary pressure–saturation relationship can be obtained via mercury porosimetry. *Acosta et al.* [3] for example performed such mercury-based measurements and transferred the results with the help of surface tensions and contact angles of air, water, and mercury to capillary pressure–saturation relationships for a water-air system. This method was first proposed by *Purcell* [84]. This approach does not take into account that gas diffusion media are mixed-wettable porous media: mercury as an almost non-wetting fluid ($\sigma_{Hg} = 140°$) does not account for the partial wettability of the carbon fibres. Hence, only one part of the p_c-S_w curve, the hydrophobic one, is represented. Moreover, the deformation of the sample under the pressure of the working fluid mercury inside the sample chamber of the mercury porosimeter is not exactly known. The reported capillary pressures of *Acosta et al.* [3] are in the range of $500\,000$ Pa up to $2\,000\,000$ Pa depending on the wettability, which is really high.

A couple of researchers proposed measurement strategies and setups to determine capillary pressure–saturation relationships for mixed-wettable gas diffusion layers. Almost all the publications are based on the principle proposed by *Bruce and Welge* [18]. *Fairweather et al.* [31], *Gostick et al.* [41], *Nguyen et al.* [71], *Cheung et al.* [23], *Harkness et al.* [44], and *Gostick et al.* [42] measured all capillary pressures depending on the saturation inside the gas diffusion media of polymer electrolyte membrane fuel cells. None of them took the exact compression level of the GDL into consideration or adjusted the compression of the fibrous porous media accurately.

Due to the mixed-wettable character of the GDL, the hydrophilic as well as the hydrophobic character has to be captured during the measurement of the p_c-S_w curves. Hence an approach with two different membranes covering the sample from both sides (cf. figure 3.2) was chosen. The break-through of air is prevented with the help of a hydrophilic membrane (GE Nylon hydrophilic supported membrane, pore size 1.2 µm, thickness ca. 110 µm), whereas the hydrophobic membrane (GE Nylon hydrophobic supported membrane, pore size 0.45 µm, thickness ca. 105 µm) ensures that no water flows into the gas port. Both membranes are highly porous (porosity in the 90 %); the influence of compression of the membranes inside the test cell during the measurements was neglected, because experiments with an incompressible porous medium (thin glass frit) and the covering membranes have shown that the influence of compression on the behaviour of the membranes is not measurable. Thus, the applied compression load to the stack of GDL and membranes is only related to the flexible, porous GDL. The

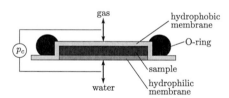

Figure 3.2: *Stack of GDL and two covering membranes [29]*

hydrophobic membrane preventing the break-through of water was tested in advance, for the purposes which a small flange with an inner diameter of 10 mm was manufactured. A circled sample of the hydrophobic membrane was mounted inside the flange and pressurized with water from one side fed via a syringe pump. The pressure itself was recorded with a pressure transducer. Via a stereo microscope the surface of the membrane was observed, and first water droplets passing through the membrane indicated the break-through pressure of the membrane. All measured break-through pressures were in the range of at least 60 kPa, which is in-line with the expected values estimated via the LAPLACE-equation.

The stack of membranes and gas diffusion media neighboured to each other is mounted inside the new measurement cell [29], sketched in detail in figure 3.3. The new design is based on decoupling the sealing unit (done with an O-ring seal) and the measurement cell where the GDL is placed and precisely compressed. The exact adjustment of the compression is necessary to mimic the loads on the gas diffusion media inside polymer electrolyte membrane fuel cells. So the compression stamp ④ sliding in a guiding bush ③ inside the upper part of the measurement cell ⑤ is moved by a linear actuator ① (Nanotec, L56, M6×0.5). To avoid distorsion of the actuator against the other parts, a distorsion lock with an internal hexagon ② is mounted additionally. The compression stamp is connected with the linear actuator via a fine pitch thread (type M6×0.5) allowing movements in micron steps. The backlash of the fine pitch thread can be accounted for via the software driving the linear actuator, but in practice the backlash was avoided by an adequate operation strategy of the testing rig, i.e. the compression stamp was retracted, subsequently the direction of movement was reversed until the front side of the compression stamp and the lower side of the upper part were in perfect alignment; then the correct overlap of the stamp was adjusted. Indicating callipers (Kaefer Feinika precision indicating calliper, resolution 1 μm, error of measurement max. 4 μm, jewel-supported bearing and impact protected) allow the positioning and checking of the upper ⑤ and lower part ⑦ of the measurement cell to each other. With the help of these callipers also the overlap of the compression stamp can be controlled. The two guiding pins ⑥, which, together with the upper part, form a high-precise fit, and the three indicating callipers help to adjust a repeatable distance of the parts to each other. Three fine pitch screws (M6×0.5) are utilized to bolt the two main parts together. The complete measurement cell is thermostated via a cryostat connected to the channels at the bottom of the lower part. The pressure difference between the water phase and the gaseous phase inside the cell is measured by a piezo-resistive pressure transducer (BD Sensors DMP 331, range of −1 to 1 bar, accuracy: $\leq 0.1\%$ FSO IEC 60770). The transducer is connected to the water inlet ⑧ by a small drilling (diameter of 1.5 mm). The pressure is measured as a relative pressure against the ambient air. The pressure of the gaseous phase inside the measurement cell is equal to the pressure outside the cell due to the hydrophobic membrane, which enables the venting of the gaseous phase, whereas the water phase is held back. This pressure difference between the two phases is also known as capillary

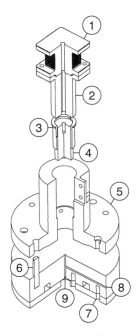

Figure 3.3: *Principle sketch of the test station [29]*

pressure (cf. equation (1.4) and figure 3.2), which is recorded with the help of LabView running on a personal computer.

The deformation of the metal diaphragm of the piezo-resistive pressure transducer under pressure may lead to impreciseness regarding the water level inside the sample, but the path length is very small (considerably less than 1 mm) and the emerging volume is negligible. Water is brought into the cell and removed afterwards via a LabView-controlled syringe pump (New Era NE-500, linked to the PC via RS232). The gas-tight micro litre syringe (Hamilton 1001 TLL) mounted on the pump is connected to the water port of the measurement cell ⑨ by a metal tubing with an inner diameter of 1 mm, which allows infusion and withdrawal of water at a broad range and in different modes (stepwise vs. continuous). Figure 3.4 shows the details of the testing rig.

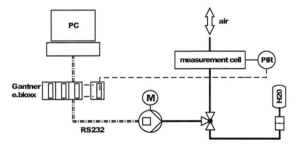

Figure 3.4: *P&ID of capillary pressure measurement*

First of all the used water in the experiments was always degassed to extract the solved air; therefore a pump with an absolute pressure of approx. 30 mbar was connected to the water reservoir. To avoid accumulation of gas inside the water-wetted parts of the measurement cell or tubing leading to inaccuracies, the complete setup was flushed with degassed water each time a new sample was mounted. The combination of syringe pump and chosen syringe provides flow rates in the range of $0.012\,\mu l/min$ up to $0.58\,ml/min$. Due to the small volumes and fluxes the deviation of the syringe pump was checked in advance: water was pumped into a glass and weighed by an analytical balance. Then the resulting flow rate was calculated. The resulting deviation was below the error in measurement.

The capillary pressure as well as the infused/withdrawn water volumes were logged on the PC and stored in a ASCII file. The pressure was recorded by I/O-modules from Gantner Instruments (e-bloxx series, 19 bit resolution) connected via RS485/RS232 to the PC. In the following the operation strategy and the necessary preparations of the samples will be described in detail.

The sample as well as the membranes were cut with hollow punches: the GDL and the hydrophilic membrane laying under the GDL sample (cf. fig. 3.2) with a diameter of 25 mm, the covering hydrophobic membrane with a diameter of 31 mm. The handling of membranes and gas diffusion media was completely done with tweezers to avoid any contamination (which may change the wettability or block pores) or damages. An O-ring sealing (nitrile butadiene rubber 70 Shore A, dimension of 25×2 mm) ensured a leak-proof sealing of the GDL and the surrounding membranes. The membranes and the O-ring sealing were replaced after each mounting of a sample.

After the sample (surrounded by the two membranes) was placed on top of the lower part above the water inlet, the upper part of the measurement cell was mounted. With the help of the adjusted overlap of the compression stamp and the indicating callipers located near the outer diameter of the cell, the compression level of the sample can be precisely set. In case of measurements without compression loads a slight compression of approximately 2 % was applied to the stack consisting of membranes and samples to ensure reliable hydraulic contact between the water-wetted parts inside the measurement cell. During the mounting processes it has to be ensured that the O-ring is placed around the compression stamp, otherwise the sample is squeezed and the sample chamber is not completely formed and filled with gas diffusion media.

If the assembly is correctly installed, the sample can be prepared by the following pre-conditioning procedure which was proposed by *Fairweather et al.* [31]. The pre-conditioning of the sample is necessary because first imbibition and drainage cycles do often not show the same behaviour as the successional cycles. The desired and measured capillary pressure–saturation curves are to reproduce the filling and draining of the pores inside the gas diffusion layers of polymer electrolyte membrane fuel cells in operation, not within the commissioning procedure.

Therefore the measurement cell is continuously filled and drained with water. As a lower limit a capillary pressure $p_{c,low} = -30\,000$ Pa was chosen, based on the assumption that all pores are empty, i.e. only the gaseous phase is present. This assumption was verified by weighing the sample with an analytical balance before the pre-conditioning and after several cycles (dismounted at the lower reversal point). The difference in weight corresponds to the residual saturation S_{wr} of the water phase. All the experiments showed that the residual saturation is always below 5 %. Based on several experiments, the weights before and after wetting the GDL, as well as both membranes were determined with a highly precise balance. For example a standard gas diffusion layer with 10 % PTFE content showed a residual saturation of 1.1 %, which is in the range of the error in measurement. A capillary pressure $p_{c,high}$ of 30 000 Pa was taken as the upper limit where almost all pores are filled with water. This value is in good agreement with the measurements of pore size distributions for GDL materials, e.g. conducted by *Ostadi et al.* [78], and the operation conditions of gas diffusion media in PEMFC. During the pre-conditioning cycles, the shape of the curves, especially the imbibition curve, change, and after several runs (normally ca. 3-5 cycles), attain a repeatable shape. The drainage curves do not change remarkably during the procedure. For details of the pressure curves during the pre-conditioning procedure refer to *Dwenger et al.* [29]. In general flow rates in the range of 3 µl/min to 10 µl/min were chosen for the pre-conditioning of the sample.

The capillary pressure–saturation curves were measured by stepwise injection and withdrawal of small portions of water. After each change of water saturation inside the sample a certain period was awaited to enable the redistribution of water. The redistribution in the pore network can be observed as a pressure change during the period. Figure 3.5 shows a typical curve where the recorded pressure is plotted against time. As the capillary pressure–saturation curves are to be measured under static conditions, the relaxation period must persist until the pressure is constant. Only when the pressure was constant, a new saturation was adjusted via the syringe pump. Normally a period of 3 to 5 min was chosen enabling the redistribution of the water phase inside the gas diffusion medium. The measured pressure at the end of the relaxation periods and the amount of infused or withdrawn water corresponds to certain points of the capillary pressure–saturation relationship. As a flow rate for the injection and withdrawal process a range between 3 µl/min and 5 µl/min was chosen, while the amount of water was between 1 µl/min and 5 µl/min. The accordant saturation level was determined

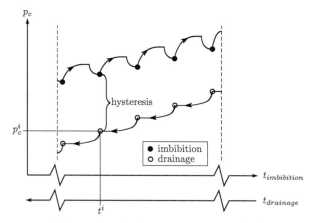

Figure 3.5: *Capillary pressure vs. time [29]*

via the pressure extrema: the saturation at each measurement point is referred to the volume of water during imbibition and drainage cycles. The capillary equilibrium existing between the covering membranes and the gas diffusion media sample, which are in hydraulic contact to each other, inside the measurement cell (due to the slow processes) allows this assumption. The infusion or withdrawal in combination with the relaxation period was repeated until the (complete) imbibition or drainage curve was measured. To this end an automated LabView program was coded enabling parallel data recording of pressure and infused/withdrawn water volume. For the analysis of the measured data a MATLAB code was developed establishing an automatized analysis of the recorded data sets.

With this setup it is also possible to measure the scanning curves for the capillary pressure–saturation relationship, which are, amongst other things, one key feature for the characterization of hysteretic porous materials. And so the main imbibition and drainage curves are determined after pre-conditioning of the sample and the membranes. Then an additional measurement cycle with adjusted infusion and withdrawal volumes is performed. A couple of these measurements allow the complete description of the area between the main imbibition and drainage curve of the p_c-S_w relationship. Again, the measured pressures can be related to the water saturation [29].

The compression level was controlled and adjusted with the help of the linear actuator and the indicating callipers. For the main purpose of the p_c-S_w measurement, the precise adjustment of the compression load onto the gas diffusion media samples was ensured by the following concept: at the beginning each material was examined without compression, i.e. the sample was mounted between the membranes and pre-conditioned (cf. also page 62). Then the desired capillary pressure–saturation curve was measured where upon the linear actuator moved the stamp downwards to the desired compression level. When the stamp was moved in the experiments, a pre-conditioning cycling was done, again guaranteeing the comparable hydraulic conditions as at the first stage. The repeat accuracy of the complete testing setup was proven with the help of multiply repeated measurements (cf. figure D.1 in the appendix). Moreover, the test equipment (syringe pump, pressure transducer etc.) were re-tested and validated monthly periods.

3.1.2 Results

In the following the results of the previous subsections will be presented. The experimentally gained data sets are in good agreement with other published measurements. *Fairweather et al.* [31] as well as *Gostick et al.* [41] have also reported maximum pressures in the range of 30 000 Pa for high saturations as well as hysteretic behaviour between the imbibition and drainage curves of gas diffusion media of polymer electrolyte membrane fuel cells.

Due to the low residual saturation ($S_{wr} < 5\%$) the plotted saturation is an effective saturation where the residual saturation is neglected. As an indication that all pores of the sample are filled and a saturation $S_w = 1$ is reached, the steep increase of the recorded pressure during the experimental determination of the p_c-S_w function is taken. As a consequence all the data are referred to a water saturation interval of $[0, 1]$.

In the following the results for various types of GDLs (cf. also appendix, page 113) are presented. At first all the results without compression are given, in parallel the influence of PTFE on the shape of the curves is highlighted, and later the compression as the most important factor is depicted.

Without compression

Figures 3.6, 3.7, and 3.8 show the capillary-pressure saturation relationship for non-treated as well as treated non-woven gas diffusion media without any compression. Additional p_c-S_w curves for SGL24BC are given in the appendix on page 144. All the materials show hysteresis between imbibition and drainage curves. As discussed on page 17, three main reasons for hysteresis in general were identified. In the present case the anisotropic distribution of the non-wetting PTFE coating all over the fibres is one reason for the hysteretic behaviour. The ink-bottle effect, driven by the shape of the pores, is the second reason for the gap between imbibition and drainage curve. In the appendix, SEM pictures of different gas diffusion layers illustrate their differences in the pore morphology and distribution of the binder and PTFE coating (cf. page 116ff). The heterogeneous character of the GDL and the very low capillary number Ca of the processes in the PEM fuel cells lead to capillary fingering effects. The very strong capillary forces inside the thin GDL effect that the water phase remains continuous, also for very low saturations effecting very low residual water saturations.

The first four materials are based on similar carbon fibre substrates leading to similar curves. The non-treated H2315 shows only small hysteresis, linked to the homogeneous contact angle of pure carbon fibre surface. In contrast, the morphology of the pores formed by the fibres lead to hysteretic behaviour. The void space defining the pores shows very different shapes if the GDL is visualised via μ-CT images. An except of the complete 3D-image is also given in the appendix for each material. If PTFE as a

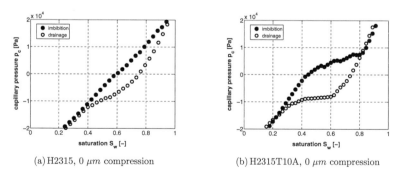

(a) H2315, 0 μm compression   (b) H2315T10A, 0 μm compression

Figure 3.6: *p_c-S_w curves of H2315 and H2315T10A*

non-wetting agent is added to the slightly hydrophilic fibrous carbon substrate (contact angle $\alpha_{carbon/water} = 86°$), the imbibition and drainage curves change (cf. H2315T10A, H2315IX53, H2315T20A). The imbibition curve of each material moves to higher capillary pressure values reflecting the hydrophobic character of the material. Comparing the imbibition curves of H2315 and the PTFE-treated materials at the same saturation, a higher pressure for wetting the pores of the hydrophobized substrates can be observed. In contrast, the drainage curves shift to lower p_c values, which has to do with the changed shape of the pores. Due to the additional PTFE, the pore diameter is decreased, which leads to higher pressure differences necessary to suck the water out (cf. also LAPLACE-equation). A similar behaviour was also measured with two

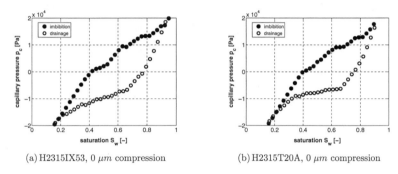

(a) H2315IX53, 0 μm compression   (b) H2315T20A, 0 μm compression

Figure 3.7: p_c-S_w curves of H2315IX53 and H2315T20A

materials (SGL24BA and SGL24BC in figure 3.8) with different morphology. The SGL materials are based on smaller fibre diameters (in the range of 2 up to 6 μm) in comparison to the other materials (fibre diameter between 9 and 12 μm). Moreover, the PTFE and the binder fixing the carbon fibres is spread all over the fibres. Now, a weaker hysteretic behaviour can be seen. The imbibition and the drainage branch of the curve are closer to each other indicating a less hydrophobic as well as hydrophilic character. The SEM pictures of SGL24BA given on page 123 suggest a more homogeneous distribution of the non-wetting PTFE inside the porous body. SGL24BC is manufactured based on the SGL24BA: a micro-porous layer is formed on top of a layer of SGL24BA. As an outcome the p_c-S_w curves look similar. Only a small jump in the imbibition curve (at water saturations S_w between 0.4 and 0.6; cf. right figure in 3.8) is linked to the additional MPL on one side of the gas diffusion layer and its wetting. For

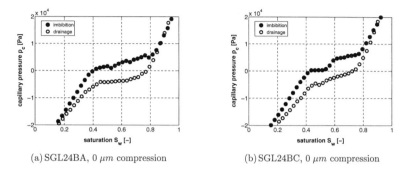

(a) SGL24BA, 0 μm compression   (b) SGL24BC, 0 μm compression

Figure 3.8: p_c-S_w curves of SGL24BA and SGL24BC

all the measurements presented here strong hysteretic behaviour was observed. For the application of these measured constitutive relationships in PEM fuel cell models, the influence of compression has to be examined. The bipolar plates compress the GDLs under the ribs and squeeze the porous media into the flown-through channels of the flow field. As a consequence, compression-depending p_c-S_w curves have to be included into the numerical models of the cathode and anode side of PEMFC. So, in the next subsections the results of capillary pressure-saturation curves for three materials precisely compressed are be given.

Influence of compression

All the described materials (cf. also overview and material characterization in the appendix) were examined with different levels of compression. Initially, the thickness of each material was precisely measured. The initial capillary pressure-saturation curve without compression was recorded, then the compression was applied in steps of 5 µm. After every single compression step the above described pre-conditioning procedure was applied. Figures 3.9, 3.10, and 3.11 show the capillary pressure-saturation curves for the compression levels of 0 µm, 10 µm, 20 µm, and 30 µm for three materials. These materials were chosen to reflect typical materials incorporated in PEM fuel cell stacks on the one hand, on the other hand the experimental validation of the complete measurements and simulations is based on a 10 % PTFE-coated gas diffusion layer of a PEMFC (cf. chapter 4).

All the measurements show the same trend: with increasing compression of the GDL, the hysteresis between the imbibition and drainage curve is getting smaller and with high levels of compression vanishes almost completely. The high porosity of the gas diffusion media in the range of 70-80 % as well as the very low residual water saturation suggest that during imbibition and drainage all the water inside the pores is connected. As an outcome, the draining and filling of the pores is not independent of each other [26]. In the previous subsection the reasons for hysteresis in GDLs are given - the surface wetting properties on the one hand, the geometry of the pores on the other hand. Due to the fact that the wettability of the gas diffusion layer is not changed substantially (the PTFE is still present) while applying the compression load onto the porous body, the diminishing hysteresis must be linked to the pore shape.

While compressing non-treated H2315, the imbibition curve does not show big changes, whereas for the PTFE-coated GDLs the shapes of the imbibition curves shift towards lower capillary pressures indicating a less hydrophobic character of the materials. Between saturations of 0.2 and 0.6 this behaviour has the biggest effect. For the drainage curve, especially at saturations in the range of $S_w = 0.4$, a trend to less hydrophilic properties arises. Due to the compression the size and the pore shape tend to get more uniform, in parallel the connectivity is enhanced further. First approaches with the virtual material design (described in the appendix page 125ff), where the granulometry method based on mathematical morphology [91] was applied, indicate such relationship. Porosimetry measurements of defined compressed gas diffusion media will help to confirm the assumption.

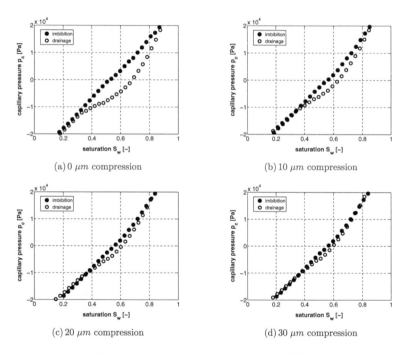

(a) 0 μm compression

(b) 10 μm compression

(c) 20 μm compression

(d) 30 μm compression

Figure 3.9: p_c-S_w curves of compressed H2315

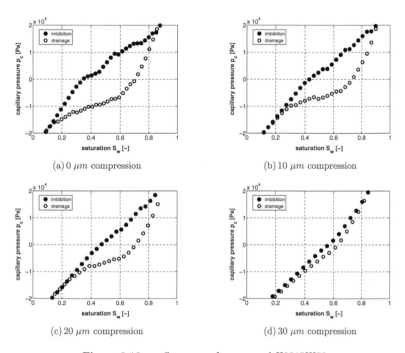

(a) 0 μm compression

(b) 10 μm compression

(c) 20 μm compression

(d) 30 μm compression

Figure 3.10: p_c-S_w *curves of compressed H2315IX53*

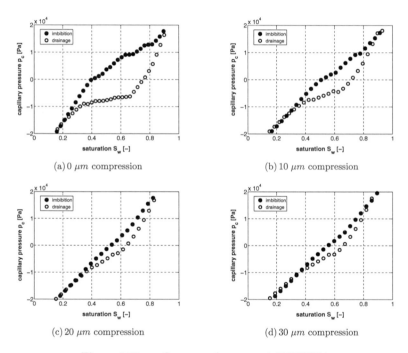

(a) 0 μm compression

(b) 10 μm compression

(c) 20 μm compression

(d) 30 μm compression

Figure 3.11: p_c-S_w curves of compressed H2315T20A

3.2 Permeability

As described at the beginning of this chapter, amongst other things the determination
of permeabilities is essential to deepen the understanding of occurring processes in fuel
cells via modelling with a multiphase Darcy approach. Moreover, the optimization of
these transport processes in the gas diffusion layers strongly depend on the material
properties and their constitutive relationships. Especially in case of gas diffusion media
as a mixed-wettable, thin technical species of porous media, the determination of constitutive relationships is a field of great interest. In the following subsections first the
experimental setup for the measurement of (relative) permeabilities is described, then
the numerical determination of k_r-$S_w$relationships is provided, and finally a summary
and comparison of results with respect to different gas diffusion layers is given.

3.2.1 Experimental setup

In the following paragraphs the experimental setups and measurement strategies for
intrinsic as well as relative permeabilities depending on the directions (in-plane vs.
through-plane) of the anisotropic fibres will be presented. Due to the anisotropic
character of gas diffusion media (cf. SEM pictures of gas diffusion media in section A
of the appendix on page 113), generally two main directions can be distinguished: in-
plane and through-plane. Through-plane means perpendicular to the fibre axis (here
z-direction, normal to catalytic layer of PEMFC), in-plane in parallel. Figure 3.12
gives a short explanation of the two directions.

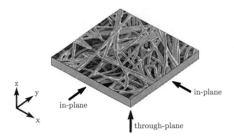

Figure 3.12: *Definition of in- and through-plane direction*

Intrinsic permeability

As a reference point for the relative permeability experiments the intrinsic permeabilities of the gas diffusion media had to be determined. Additionally, different levels of compression were applied to the examined structures. Steps of $5\,\mu m$ displacement beginning without compression ($0\,\mu m$) up to $75\,\mu m$ compression of the GDLs were chosen. The measurement cells, designed for the measurement of relative permeabilities, were operated for these tests in single-phase mode, i.e. the water phase port was removed and only the gaseous phase was fed. The pressure drop between the inlet and the outlet of the cell was measured and recorded. Initially gas flux through the empty cell was varied and the overall pressure drop of the device itself was measured. Subsequently, the overall pressure drop was subtracted from the resulting pressure drops of all measurements with GDLs. Moreover, the resulting (intrinsic) permeabilities of gas diffusion media were taken as a reference for the determination of relative permeability.

The experimental determination of permeabilities for gas diffusion media (non-woven as well as woven structures) of polymer electrolyte membrane fuel cells was described by several authors [20, 32, 39]. In general the proposed techniques for gas diffusion media descend from soil–physics motivated standards, which are for example given by the DIN or ASTM International [6–8, 25]. The permeability K is linked to Darcy's law (given in equation (1.15)), which leads to the volume flux related Darcy's law by multiplying the term and the flown-through area A:

$$Q = A \cdot \frac{K}{\eta} \cdot \frac{\Delta p}{d} \tag{3.1}$$

If the flux of species is increased, a non-linear dependency of the pressure drop of the velocity / volumetric flux is observed due to turbulence inside the flown-through porous body. These effects can be described with the help of the Forchheimer equation. In all conducted experiments it was ensured that the Reynolds numbers Re were small enough, which leads to the applicability of Darcy's law — here Forchheimer effects are not present.

On page 22 the equation for the relative permeability and its derivation are given. Because of the anisotropic character of the gas diffusion layers, two main directions were regarded for the determination of relative permeabilities in general. Depending on the direction of flux (cf. figure 3.12), area A of the sample changes. For through-plane permeabilities equation (3.1) can easily be applied, whereas for in-plane permeabilities due to varying radius r of the sample (cf. also figure 3.13), the equation (3.1) cannot be applied. In the following, a brief description of the deviation of equations for in-plane

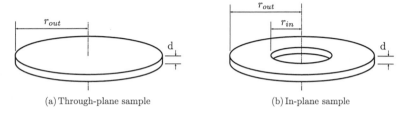

(a) Through-plane sample   (b) In-plane sample

Figure 3.13: *Shape of samples*

measurements is given, wherein mass conservation for the domain is ensured by the following equation:

$$\frac{d}{dr}(\rho \cdot v \cdot A) = 0 \tag{3.2}$$

While applying equation (1.15) and equation (3.2) as well as the ideal gas law, the mass conservation equation reads as follows:

$$\frac{d}{dr}\left(\frac{p}{R \cdot T} \cdot \frac{K}{\eta} \cdot \frac{dp}{dr} \cdot (2 \cdot \pi \cdot r \cdot d)\right) = 0 \tag{3.3}$$

The pressure boundary conditions for the in-plane case (cf. also figure 3.13(b)),at the inner position r_{in} and at the outer one r_{out}, are given by:

$$p(r) = \begin{cases} p_{in} & \text{for } r = r_{in} \\ p_{out} & \text{for } r = r_{out} \end{cases} \tag{3.4}$$

where p_{in} and p_{out} are given in absolute pressures. Moreover, assuming isothermal conditions and neglecting heat transfer from the gas flux to the porous medium and vice versa, the integration of equation (3.3) leads to:

$$r \cdot p \cdot \frac{dp}{dr} = c_1 \qquad (3.5)$$

Repeated integration of equation (3.5) gives:

$$\frac{p_{out}^2 - p_{in}^2}{2 \cdot ln\left(\frac{r_{out}}{r_{in}}\right)} = c_1 \qquad (3.6)$$

The combination of equations (3.3) and (3.6) leads to the final equation used for the evaluation of the in-plane permeability of gas diffusion media:

$$\frac{dp}{dr} = \frac{1}{r \cdot p} \cdot \frac{p_{out}^2 - p_{in}^2}{2 \cdot ln\left(r_{out}/r_{in}\right)} \qquad (3.7)$$

As a basis for the evaluation of the permeability of the gas diffusion media the pressure and the gas flux were taken. Hence equation (3.7) can be reformulated in combination with Darcy's law as:

$$K = ln\left(\frac{r_{out}}{r_{in}}\right) \cdot \frac{\eta \cdot p_{out} \cdot Q}{2 \cdot \pi \cdot d \cdot (p_{in}^2 - p_{out}^2)} \qquad (3.8)$$

The set-up and the equipment for measurements is described in the following subsection. The testing rigs for the measurements of in-plane as well as through-plane relative permeabilities can also be taken for the measurements of intrinsic permeabilities. Here, only single-phase flow is applied and the resulting pressure drop is taken to calculate the intrinsic permeability of the porous gas diffusion medium, based on the equations described above.

Relative permeability

In contrast to intrinsic permeability relative permeability depends on several properties: again, the structure of the porous medium, but also on the interactions between the two mobile phases (strongly depending on the saturation level inside the porous medium). Moreover, the interactions with the pore structure have an influence on the relative permeability itself. The relative permeability as one crucial relationship for the description of transport processes inside porous media such as mixed-wettable porous backings, which, in the present case, consist of gas diffusion layers of polymer electrolyte membrane fuel cells, can be determined with the help of numerical techniques [14, 45] or experiments.

In general the experimental determination of relative permeabilities can be subdivided into two fields: the unsteady flow methods and the steady flow techniques. The unsteady flow methods are characterized by the displacement of one phase through the second one. *Welge* [98] has extended the Buckley-Leverett frontal advance equation to derive values for relative permeabilities. Thereby the relative permeability is calculated from the ratio of outflowing fluids in the experiment. The main plus of the unsteady method is speed and the possibility of choosing flow conditions identically to the real scale problem. Several researchers have shown that the results of unsteady methods in general are in good agreement with those of steady state experiments, but numerous exceptions exist [26]. The main disadvantage of the unsteady experiment stems from the treatment of boundary effects: elevated pressure differences between the two phases must be applied to avoid them at the outlet. Moreover, high (water) saturations can not be realized without changing the fluid system (e.g. water is replaced by high viscous oil). Otherwise, at high flow rates, water is pushed out of the sample like a piston - consequently, the derived k_r-S_w curves are limited to a small number of working conditions.

While measuring relative permeabilities with steady state methods, both fluids are injected with continuous and constant flow rates. They are varied in different experiments to change the saturation inside the porous sample. For each run the pressure drop across the sample is measured and recorded. The flow conditions are not changed until the pressure drop is constant and the inflow rate is equal to the outflow rate, which ensures that the saturation inside the sample does not change anymore. Thereby a uniform saturation inside the sample without any gradients is assumed. Mostly, the pressure drop is measured only in the fluid phase with the higher pressure: the expectation that the pressure drop in the second fluid is the same is based on the supposition that the saturations are all over the sample are equal to each other.

As always the relative permeabilities are related to the saturation of one phase. In the present case the water saturation is chosen. For the exact value of saturation the amount of water inside the test sample must be identified. There are several possibilities: determination via weighing, measurement of the electrical resistivity of the sample, volumetric balance of gas and water phase, and nuclear magnetic resonance or computer tomography via x-rays or neutron radiation of the measured sample in combination with image analysis techniques. To simplify matters, a gravimetric analysis of water saturation was chosen for the experiments as described in the following.

In the literature a couple of measurement setups and methods are described: the Hassler method, the Hafford technique (simplification of the Hassler method in the case of high flow rates where capillary end effects are avoided: the diaphragm is omitted, cf. also *Osoba et al.* [77]), the Penn State method [68], single-core dynamic, the gas-drive method, and the dispersed feed setup. For more details refer to *Scheidegger* [85], who gives a comprehensive summary in tabular form.

In the case under consideration, the Penn State method as a steady-state technique was chosen because boundary effects (gradients in saturation or pressure at the beginning or at the end of the sample) can be eliminated reliably. To eliminate these effects, a mixing and an end section are required. Figure 3.14 shows the k_r-S_w measurement cell for through-plane measurements. As previously described water and air are injected via ① into the measurement cell. The premixing section ② helps to build a well-mixed gas/water feed, i.e. both phases are equally distributed across the cross section of the inlet. So a porous plastic body with an average pore diameter of 80 µm was taken. The alignment pins ④ help to position both parts of the measurement cell accurately to each other. Moreover, the distance between the two aluminium machined parts is measured with the help of (in total) three indicating callipers, which, for example are placed at position ③. In general the test cells were manufactured with regard to high quality surfaces, plane parallelism, and almost free from unwanted play, which leads to very small production tolerances.

The gas diffusion layer sample is placed between the two halves of the cell (position ⑤). Around the sample, an O-ring seal is placed, which ensures leak-free operation of the cell. Additionally, the porous plastic parts ② and ⑦ are sealed at their outer diameter with O-rings to avoid by-pass flows. The brass thread insert ⑥ was incorporated as counterpart of the three fine pitch thread screws (M6 x 0.5) fixing both major parts and helping to adjust the correct distance between the two halves of the measurement cell. As an end section ⑦ again porous plastics with properties as described above were chosen. The outlet ⑧ completes the setup. The pressure drop of the (water-filled) sample is measured between the inlet and the outlet. As a reference state the pressure drop of the cell without any sample is measured in advance.

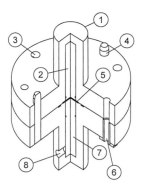

Figure 3.14: *Principle sketch of the k_r-S_w through-plane measurement cell*

The cell for the in-plane relative permeability measurements is similar to the through-plane cell. In figure 3.15 a sketch of the cell is given. The principle as described above, with an inlet on top ①, a premixing section ②, positions for the indicating callipers ③, and guiding pins ④, is applied. In contrast to figure 3.14, where the sample is axially flown-through, the flux from the premixing section crosses sample ⑤ in radial direction. The collecting cut-in ⑥ is connected with the outlet ⑦ and canalizes the air/water mixture. Identically to the through-plane k_r-S_w measurements, the pressure drop of the empty cell is determined in a first step. With both the measurement cells the isolation of in- and through-plane relative permeabilities under various compression levels is possible. The complete experimental setup consisted of several LabView-controlled components (cf. also figure 3.16 for details of the P&ID). Gantner Instruments e.bloxx I/O-modules, which were connected to a PC via a RS232-RS485 converter, were chosen to control the testing rig. The gas flow was adjusted with the help of a MKS Instruments mass flow controller (type 1179A) with a range of 0 up to 500 sccm, controlled by a controller unit (MKS 647B). The controller unit was also connected with the PC via a RS232 port. Moreover, water was fed via a HPLC pump (Merck Hitachi L6200 Intelligent Pump) with a range of $0^{\text{ml}}/\text{min}$ to $9.999^{\text{ml}}/\text{min}$. The resulting

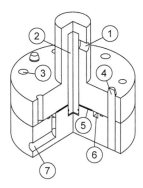

Figure 3.15: *Principle sketch of the k_r-S_w in-plane measurement cell*

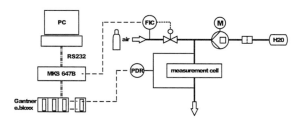

Figure 3.16: *P&ID of relative permeability measurement*

water saturation S_w inside the sample was gravimetrically measured with an analytical balance manufactured by Kern & Sohn GmbH (type ABJ 220-4M; capacity 220 g, readability 0.1 mg, repeatability 0.1 mg, linearity 0.2 mg). The pressure drop over the (water-filled) sample was detected by a pressure difference gauge (JUMO GmbH & Co. KG, type 4304, $p_{max} = 50$ mbar, measurement error in sum below 1 %). As described in section 3.1, the compression of the samples was adjusted with indicating callipers (for technical specifications cf. page 59).

As stressed above, the boundary effects must be avoided by all means while measuring k_r-S_w relationships for GDLs under various loads of compression. The method (also known as the steady state displacement method) itself was mainly developed by Yuster and co-workers [68] at the Pennsylvania State University and widely applied to groundwater or reservoir engineering problems. The main advantage of the Penn State method is the applicability to imbibition and drainage as well as the above mentioned avoidance of boundary effects, and the possibility to decompose easily the measurement cell for saturation determination via the weighing of the sample.

In the preceding paragraph it was shown that the measurement of relative permeability saturation relationships for gas diffusion layers of fuel cells is a challenging task. The thickness of the GDLs, which is approximately in the range of 200 up to 300 μm, leads to the requirement of precise specification of compression inside the measurement cells. The constant supply of low mass flows for the water and the gas phase and the pre-mixing inside the measurement cell are also critical tasks. The accurate determination of the pressure drop across the sample is required to evaluate the corresponding relative permeability correctly.

The results of the k_r-S_w measurements are utilized later in the simulation of counter-current transport of gas and water in thin mixed-wettable porous gas diffusion layers. With this aim in view the assumptions of the taken framework of equations must be held. Especially Darcy's law is based on the creeping flow of species. As described in chapter 1 (cf. equation (1.1) on page 16), the dimensionless capillary number Ca as one criterion and the Reynolds number Re (cf. equation (1.13) on page 21) were taken. In the appendix on page 141 the estimations of present Re- and Ca-numbers are given and the validity of Darcy's law is shown.

3.2.2 Numerical determination

For the numerical determination of k_r-S_w relationships for GDLs several techniques and ways are described in the literature *Koido et al.* [57], *Shia et al.* [87], and *He et al.* [45], for example, have presented different approaches. Based on the results obtained by the stochastic optimization of interfacial energies (cf. chapter 2), relative permeabilities for virtually generated gas diffusion media are derived. The determination of (relative) permeabilities depending on the direction which builds the permeability tensor (cf. equation (1.16) on page 21) is a main advantage of the presented technique. Figure 3.1 depicts the general approach: at the beginning the desired structure is generated with *genGDL* for non-woven and *genCDL* for woven structures. Details are given in section B of the appendix. The simulated annealing approach (cf. 31ff.) is chosen to search the energetic minimum for a predetermined water saturation. The interface between the gas phase and all other phases (solid, PTFE, and water) is adjacently detected. With the help of a tesselation of the resulting interface a stl-file is generated. The stl-format was chosen by virtue of simplicity and popularity. Meshing and preprocessing tools (in the present case *ANSYS ICEM CFD*) provide a tetrahedral mesh based on the stl-file, which is embedded in a pre-defined modelling domain.

In the final step the resulting mesh enables the solution of the incompressible, isothermal Navier-Stokes equation (cf. equations (3.9), (3.10) in the following) for the given structure with a defined saturation.

continuity:

$$\nabla v = 0 \qquad (3.9)$$

momentum:

$$\rho \cdot \left(\frac{\partial v}{\partial t} + (v \cdot \nabla) \, v \right) = -\nabla p + \eta \cdot \Delta v + f \qquad (3.10)$$

If the body forces f are neglected, the 3-D incompressible Navier-Stokes equations in Cartesian coordinates read as follows.

continuity:

$$\frac{\partial v_x}{\partial x} + \frac{\partial v_y}{\partial y} + \frac{\partial v_z}{\partial z} = 0 \tag{3.11}$$

momentum in x-direction:

$$\frac{\partial v_x}{\partial t} + v_x \cdot \frac{\partial v_x}{\partial x} + v_y \cdot \frac{\partial v_x}{\partial y} + v_z \cdot \frac{\partial v_x}{\partial z} = -\frac{1}{\rho} \cdot \frac{\partial p}{\partial x} + \nu \cdot \left(\frac{\partial^2 v_x}{\partial x^2} + \frac{\partial^2 v_x}{\partial y^2} + \frac{\partial^2 v_x}{\partial z^2} \right) \tag{3.12}$$

momentum in y-direction:

$$\frac{\partial v_y}{\partial t} + v_x \cdot \frac{\partial v_y}{\partial x} + v_y \cdot \frac{\partial v_y}{\partial y} + v_z \cdot \frac{\partial v_y}{\partial z} = -\frac{1}{\rho} \cdot \frac{\partial p}{\partial y} + \nu \cdot \left(\frac{\partial^2 v_y}{\partial x^2} + \frac{\partial^2 v_y}{\partial y^2} + \frac{\partial^2 v_y}{\partial z^2} \right) \tag{3.13}$$

momentum in z-direction:

$$\frac{\partial v_z}{\partial t} + v_x \cdot \frac{\partial v_z}{\partial x} + v_y \cdot \frac{\partial v_z}{\partial y} + v_z \cdot \frac{\partial v_z}{\partial z} = -\frac{1}{\rho} \cdot \frac{\partial p}{\partial z} + \nu \cdot \left(\frac{\partial^2 v_z}{\partial x^2} + \frac{\partial^2 v_z}{\partial y^2} + \frac{\partial^2 v_z}{\partial z^2} \right) \tag{3.14}$$

$Open\nabla FOAM$ [76] provides with *icofoam* a transient toolbox solving the incompressible Navier-Stokes equations (3.11), (3.12), (3.13), and (3.14) using the PISO (**P**ressure **I**mplicit with **S**plitting of **O**perators) algorithm. The PISO algorithm is a further development of the SIMPLE algorithm.

The complete virtual porous medium is located between in- and outlet volumes, which are both neighboured to the structure. Figure 3.17 gives an impression of a flown-through virtually generated porous gas diffusion layer without water inside. For both the additional volumes at least volumes of the same length as the virtual GDL before and after the porous body were chosen. At the inlet a Neumann boundary condition was chosen, at the outlet the pressure was set to a defined value (Dirichlet boundary condition). The inlet normal velocity was chosen in a way that the corresponding pressure drop of the porous structure was in the range of 200 Pa ensuring laminar conditions. The walls and the surface of the structure also have a Neumann boundary condition. The relative permeability is finally calculated with the help of equation (1.18). To this end, the averaged pressures at the inlet and the outlet (which are already given at the outlet by the Dirichlet boundary condition) will be taken.

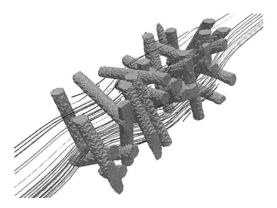

Figure 3.17: *Flown-through structure with streamlines*

3.2.3 Results of k_r-S_w relationships

In the following, permeabilities for the in- and through-plane directions are given. Various materials with different PTFE content were tested and modelled. At the beginning of the second subsection the intrinsic permeabilities are presented, followed by the relative ones.

Here, three types of gas diffusion media were chosen for the simulations: without PTFE, with five percent, and with a 10 % hydrophobic PTFE coating. Additional results can be found in the appendix on page 143.

Typically the relative permeabilities depend on the kind of displacement: in the case of imbibition the distribution of the phases is different in comparison to the drainage, which leads to two branches of curves for each process. In the case of the performed simulations, the water phase as non-wetting phase was fixed a priori. Hence a differentiation between drainage and imbibition cannot be performed. The experimental determination (with the Penn State method) of relative permeabilities for the water phase is described above. Due to the low water saturations, different relative permeabilities depending on wetting or draining the porous medium were not observed.

Simulations

The simulations were performed on the basis of the virtually generated structures utilized during the optimization process for the water distribution. The two major directions of flow through GDLs were examined. In the following the relative permeabilities
of the gaseous phase are plotted versus the water saturation inside the structure. The

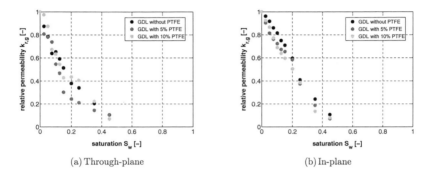

(a) Through-plane   (b) In-plane

Figure 3.18: *Comparison of relative permeabilities depending on hydrophobic coating*

relative permeabilities drop down towards zero due to the blocking effect of the accumulating water inside the fibrous porous media. On page 44ff the blocking of areas
inside the GDL can easily be seen. The amount of PTFE causing the mixed-wettable
behaviour has no significant influence on the shape of the curve. Structural effects
are dominating. Moreover, the anisotropic characteristics of the gas diffusion media
are responsible for the different shape of the curves: in the through-plane direction
the relative permeability is dropping down very fast, the in-plane directions show an
almost linear behaviour. At saturations of 0.2 the value for $k_{r,g}$ is in the range of 0.4
for the through-plane direction, whereas in the in-plane direction the values for the
relative permeabilities are approximately 0.6.

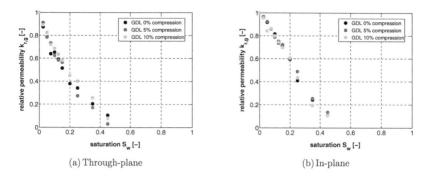

(a) Through-plane   (b) In-plane

Figure 3.19: *Comparison of $k_{r,g}$ of untreated GDL under compression*

Experiments

The experiments were conducted with the testing rigs shown already. Again, in- and through-plane directions were measured resulting in the constitutive relative permeability saturation relationship as depicted in the figures below. Moreover different levels of compression were applied to the gas diffusion media.

The intrinsic permeabilities given in figures 3.20(a) and 3.20(b) as well as figures D.2 - D.5 in the appendix were taken as a basis for the determination of the relative permeabilities. In the in-plane direction the permeability is approximately one order larger owing to the anisotropic character of the gas diffusion media. Moreover, the amount of hydrophobic PTFE plays an important role: all the more as the permeability declines, which is due to the decreasing pore space inside the GDLs. If the porous diffusion layers are compressed, the flown-through space is also reduced, leading to smaller permeability. In a subsequent step the relative permeabilities were measured with the help of the measurement cells and strategies described above. Again, in-plane and through-plane directions are plotted separately for three different types of materials: gas diffusion layer without PTFE, one with 10% PTFE, and a diffusion medium with an extra micro porous layer. In the through-plane direction the relative permeabilities for the water phase strongly depend on the material properties. The non-treated material (black markers in figure 3.21(a)) shows small values in comparison to the two other materials with PTFE. One explanation for these results is the distribution of

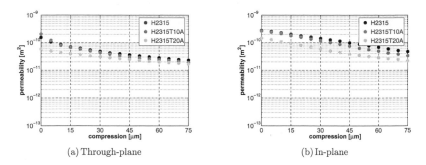

(a) Through-plane   (b) In-plane

Figure 3.20: *Comparison of intrinsic permeabilities under compression*

the water phase inside the porous GDL (for details cf. chapter 2): the GDL without
a surface coating has a slightly hydrophilic behaviour enhancing the formation of a
water film on the surface of the carbon fibres, whereas the two other hydrophobic materials possessing a minimized non-treated surface show only few sites of pure carbon
fibres. In general, these sites are preferentially wetted by the water phase. Hence, the

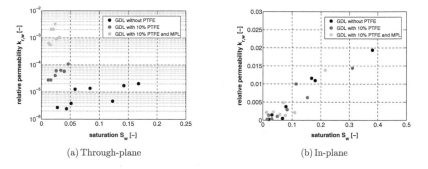

(a) Through-plane   (b) In-plane

Figure 3.21: *Comparison of relative permeabilities*

conductivity for the water phase, which plays the most important role for the relative
permeability, strongly depends on the water distribution inside the porous medium.
In the case of the GDL with PTFE coating and the type additionally equipped with
a MPL a small number of hydrophilic sites form a flow path for the water phase. In
contrast, the non-treated surface of the GDL without PTFE tends to form a water

film on the surfaces of the fibres. For the in-plane direction the influence seems to be negligible, which is linked to the anisotropic character of the gas diffusion media. In that case fibres are orientated in the flow direction enhancing the transport of the water phase.

3.3 (Effective) Diffusivity

The diffusivity of gases inside the gas diffusion layer strongly determines the transport of reactants to the reaction sites. Resulting reaction rates, which are directly coupled with the power output of the fuel cell, mainly depend on these transport phenomena. A modelling approach describing the impact of saturation onto the diffusion is presented in the following and the structural influence will be highlighted. Additionally, the measurement and analysis of these diffusivities for thin mixed-wettable porous layers under various compression levels will be described in detail.

Several authors have already measured diffusion coefficients for a couple of types of GDL with different techniques. *Sunakawa et al.* [94] measured diffusivities of a complete fuel cell. *Flückiger et al.* [33] applied the electrochemical diffusimetry method to determine the effective diffusivity depending on PTFE content and direction. *Zamel et al.* [104] chose a Loschmidt diffusion cell to measure the effective diffusion coefficient of oxygen-nitrogen mixtures depending on temperature, PTFE treatment, and porosity. *LaManna and Kandlikar* [58] measured the water vapour diffusion coefficient with the help of a parallel flow mass exchanger where a GDL was placed as a permeable separator in the middle. The influence of MPL, thickness, and PTFE loading was examined.

In almost all the examinations cited above the most familiar description of the diffusion process is Fick's law. Fick's law is a simplification of the Stefan-Maxwell approach, so infinite dilution, constant temperature, and ideal behaviour of the considered phase are assumed. Hence, the first Fickian law says that the diffusive flux is proportional to the gradient of concentrations in space:

$$J = -D \cdot \frac{\partial c}{\partial x} \qquad (3.15)$$

With the help of equation (3.15), the effective diffusivity through a porous medium (with cross-section A and thickness d) as the ratio of the diffusion flux through a porous medium related to the undisturbed diffusive flux (bulk diffusion), can be written as:

$$D_{eff} = \frac{J_{porous}}{J} \tag{3.16}$$

In the experiments, which will be described in the next subsection, the gradient Δc_{porous} over the sample was measured. If equation (3.15) is applied to equation (3.16) the molar flux J can be expressed as concentration gradient and the relative diffusivity reads as:

$$D_{eff} = \frac{\Delta c_{porous}}{\Delta c} \tag{3.17}$$

If the component i is balanced in a differential volume element and the first Fickian law is applied, the second Fickian law describing a non-steady diffusion process will arise. The time-dependent concentration gradient is given in the following.

$$\frac{\partial c}{\partial t} = D \cdot \left(\frac{\partial^2 c}{\partial x^2} + \frac{\partial^2 c}{\partial y^2} + \frac{\partial^2 c}{\partial z^2} \right) \tag{3.18}$$

3.3.1 Determination with Wicke-Kallenbach cell

The setup for the determination of effective diffusion coefficients is often based on the Wicke-Kallenbach cell. *Keil* [53] gives a good overview about diffusion phenomena inside porous media and the basics of experimental determination of diffusion coefficients. *Park et al.* [81] have reviewed the measurement of effective diffusivities in porous media in a comprehensive way. Figure 3.22 depicts the main principle of the Wicke-Kallenbach measurement cell. Both sides of the cell are flown through by gases (labelled with A and B) and held at the same temperature level. The mixture of gases (given as A(B) and B(A) respectively) leave the cell and are then analysed. All measurements were done as stationary experiments. In the present case the pressures in both chambers of the cell were the same all over the time so that only diffusion processes through the sample in between (depicted here in dark-grey) occurred. The sample was a cut-out of a GDL sheet with a hollow punch and mounted carefully. The

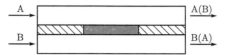

Figure 3.22: *Principle sketch of a Wicke-Kallenbach cell*

fibrous gas diffusion media of polymer electrolyte membrane fuel cells are anisotropic materials, hence the main two directions (in-plane vs. through-plane, cf. also figure 3.12 on page 73) are incorporated in the designs of the measurement cells. Figure 3.23 and 3.24 show cross-sections of both cells. The through-plane measurement cell consists mainly of two parts where each has a small plate supporting the sample (cf. small hatching in figure 3.23). The sample lies between the two plates (given in light-grey in both figures) and has circular shape, cut with the help of a hollow punch. Inside each plate (diameter of 47 mm) 45 precisely drilled holes (diameter of 1.5 mm) can be found. These holes are positioned vice versa in each plate enabling a diffusive flux through the sample. Small pins act as distorsion lock for the plates. The compression level of

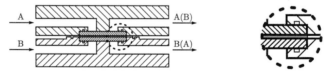

Figure 3.23: *Principle sketch of through-plane cell*

the sample is set via fine-pitch threaded screws and controlled by indicating callipers (cf. page 59 for details of the callipers). The complete setup is thermostatized with the help of the lower part owning machined cooling ribs, which are connected to a cryostat, and the temperature of the cell and the fed gases are controlled via thermocouples. The different parts of the cell are sealed against each other by O-ring seals. Again, the O-ring around the sample was replaced after each measurement.

The in-plane measurement cell has a similar shape – two main parts are flown through and the sample itself is placed between them. O-rings ensured a proper sealing of the cell. The distance the two aluminium parts were adjusted in the same way as described above. Additionally the temperature of the cell is controlled via thermocouples and

kept constant by a cryostat. The sample has the shape of a ring (light-grey in figure
3.24) and was punched with a proprietary hollow punch. The peripheral devices are

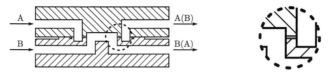

Figure 3.24: *Principle sketch of in-plane cell*

the same for both measurement directions; in figure 3.25 the P&ID of diffusivity measurement is given. The complete testing setup was controlled by a self-coded LabView
program: Gantner Instruments e.bloxx modules (A3-1, A4-1TC, and A9) were connected with the PC via RS485/RS232. The gas flow was adjusted with the help of a
variety of mass flow controllers (MKS, type 1179A or Bronkhorst EL-flow select) with
different maximum flow rates: 75 sccm, 200 sccm for the sample gas flowing through
the cell, and 500 sccm or 1000 sccm for test gas, which was used for the calibration
of the mass spectrometer (cf. also P&ID in figure 3.25). Nitrogen and argon were
chosen as gases because of their inert behaviour and their low adsorption power. The
gas flow was adjusted by potentiometers controlled via the e.bloxx modules and Lab-
View. All the mass flow controllers were continuously calibrated with a DryCal Definer

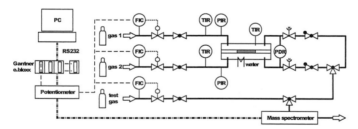

Figure 3.25: *P&ID of diffusivity measurement*

gas flow meter (Bios International). The exact pressure level for the two sides of the
measurement cell were regulated with fine pitch thread needle valves. Additionally,
the pressure drop across the sample was controlled with a micromanometer (Furness
FCO12, range of 0 to 2 mbar or up to 20 mbar). The absolute pressure for all gas

streams are measured and recorded with precise pressure transducers (BD sensors, Wika). The exact compression level of the gas diffusion layers mounted inside the measurement cells were checked and adjusted by the already described Kaefer Feinika precision indicating callipers. The resulting gas concentrations were measured with a mass spectrometer (Balzers Instruments, ThermoStar GD 300 T2, Quadrupole mass spectrometer, QMS 200 M1 analysator), which was re-calibrated in advance before each measurement. For the calibration itself, test gases with a precisely known concentration were used. The measured concentrations were also recorded as an ASCII file enabling an analysis and visualization of the complete recorded data with a self-coded MATLAB script.

Due to the stationary measurement principle, the concentration profile for each direction is constant after some time. Therefore the general mass balance reads as (here given for the z-direction):

$$\frac{\partial^2 c}{\partial z^2} = 0 \tag{3.19}$$

The following boundary conditions are valid for the setup described above under the assumption of a continuously stirred tank reactor at each side of the sample:

$$\text{at} \quad z = 0 \quad \text{with} \quad c = c_1 \qquad \text{and at} \quad z = d \quad \text{with} \quad c = c_2 \tag{3.20}$$

Hence, the molar flux in the considered direction of the porous gas diffusion medium is given by (cf. also equation (3.15)):

$$J_{porous} = D \cdot \frac{c_1 - c_2}{d} \tag{3.21}$$

The effective diffusion coefficient as the ratio of bulk diffusion and diffusion through the porous medium can easily be obtained from the measured concentrations and the belonging diffusion coefficient in the bulk phase (cf. equation (3.16)). The bulk diffusion coefficient for argon / nitrogen were estimated with the help of the empirical correlation proposed by *Fuller et al.* [34], [35], [36]. The details of the correlation are given in the appendix on page 142.

3.3.2 Numerical approach

Several researchers have proposed models for the prediction of diffusivities in porous media. Again, similar to the previous section, the direction-depending transport properties of diffusivities can be derived and expressed as a tensor. In general, the effective medium theory is well known in the literature. Especially for gas diffusion media of polymer electrolyte membrane fuel cells, a series of publications with different approaches can be found. *Nam and Kaviany* [70] have modelled the GDL as intersecting fibres building regular squared pore spaces. Based on this model they derived in- and through-plane diffusivities, which were later used in a 1D-model of a complete PEMFC. *Das et al.* [24] have formulated a model based on the Hashin Coated Sphere model to avoid the usage of several empirical correlations in parallel.

Wu et al. [102] have proposed a three-dimensional pore network model where the GDL is represented as a regular network built by spheres and cylinders mimicking pores and throats. Again, Fick's law was applied to calculate the oxygen effective diffusivity. The effect of network size, heterogeneity, pore connectivity, and anisotropy was investigated. Later on the same group [101] modelled the effective diffusivity of oxygen via a fractal model. Mixed wettability, tortuosity, and the Knudsen effect were also incorporated into the model.

In the present case, the results of chapter 2, i.e. the water distribution inside the virtually generated porous medium, were chosen as a basis for further calculations. Figure 3.1 on page 54 shows the general way for the determination of transport parameters. The diffusion process is described with Fick's law (cf. also equation (3.15) and (3.18) at the beginning of the section).

For the solution of the boundary value problem OpenFOAM [76] was used. OpenFOAM provides several pre-defined solvers for a couple of problems. In the present case the solver *laplacianFoam* was chosen to solve the posed diffusion equation. Neumann boundary conditions were chosen for the outlet, the walls, and the surface of the structure. A Dirichlet boundary condition was taken to represent the inlet concentration. The concentration all over the inlet was the same and was chosen in the range of $0.01\,^{mol}/m^3$ to avoid non-physical high molar fluxes let the assumptions for Fick's law might be violated. Moreover, such high molar fluxes need very fine meshes leading

to high computational demand and numerical challenges. Figure 3.26 illustrates the
principle approach and the resulting modelling domain. As initial condition one con-

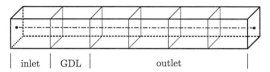

| inlet | GDL | outlet |

Figure 3.26: *Modelling domain with in- and outlet volumes*

centration of the two species is set equal to zero. Hence, a diffusive flux through the
domain is established when the simulation is started.

Additional in- and outlet volumes are neighboured to the virtual porous medium to
avoid effects originating from the boundaries. These effects at the inlet will arise due to
the concentration front migrating through the domain and accumulating in front of the
porous medium because of the reduced cross section. At the outlet side of the virtual
(water-filled) GDL, the domain is extended to avoid back-diffusion from the outlet into
the porous domain where the effective diffusion coefficient is evaluated. The length
of the additional inlet volume is the same as the virtually generated GDL, the outlet
volume is four times larger than the domain of the porous medium.

The analysis of the simulation data and the determination of the effective diffusion coefficient was performed with ParaView [80]. And so the results file with time-dependent
concentration information was loaded and a middle axis (dashed line in figure 3.26)
in the direction of interest was set. Perpendicular to this axis a cut plane, each at
the beginning and at the end of the virtual GDL, was established. The concentration
gradients in each spatial direction were evaluated and the area-averaged concentrations
c_1 and c_2 at these cut planes were determined via an integral.

In figure 3.27 an example with a water-filled structure ($S_w = 0.15$) is given. The inlet
is in front of the picture and the above described boundary conditions are applied.
The gradients except the gradient in the considered direction have to vanish, otherwise
due to local concentration gradients the result is misleading. The concentrations c_1
and c_2, which result from the area averaging are taken to calculate the molar flux
J_{porous} through the porous medium. With the above described technique effective diffusion coefficients for almost all porous media and for every spatial direction can be
determined.

Figure 3.27: *Partly water-filled structure with meshed domain*

3.3.3 Results

In the following paragraphs the effective diffusivities of several gas diffusion layers are presented. On the one hand experimentally determined results are given, on the other hand the results from the second chapter leading to partially water-filled structures, which were again used for the numerical determination, are shown. Exemplarily the simulations of a blank GDL without PTFE, a gas diffusion medium with 5 %, and a species with 10 % are given. Moreover, different levels of compression are applied to the virtually generated porous media.

Simulations

The following results were derived according to the approach described at the beginning of the present chapter. In-plane as well as through-plane diffusivities for different structures and saturations are given. It can clearly be seen that there is a strong dependency of the diffusivities on the present water saturation. The structures with five and ten percent PTFE show smaller values because of reduced void space inside the GDL. Moreover, compression of the virtual structures leads to less void space, which is again the reason for worse properties in the context of fuel cells. While the water saturations are increased, the through-plane calculations only show a small reduction of the diffusivity whereas the in-plane cases show a dramatic descent of the values. This behaviour originates from the strongly anisotropic properties of the gas diffusion media:

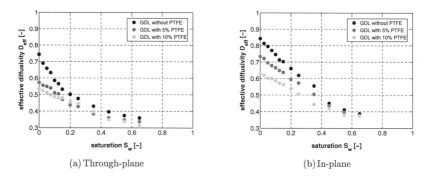

(a) Through-plane   (b) In-plane

Figure 3.28: *Comparison of calculated effective diffusivities depending on PTFE content*

almost all the fibres are orientated in the x-y plane. In a second step the influence of

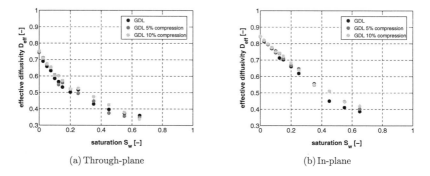

(a) Through-plane   (b) In-plane

Figure 3.29: *Calculated effective diffusivity of untreated GDL under compression*

compression on the effective diffusivities was examined with the help of three stages of compression (cf. also page 125ff for details) showing that the effect is insignificant. Referring the effective diffusivities to the saturation, i.e. directly depending on the void space inside the porous gas diffusion layer as a compression-independent quantity, allows a certain portability to various compression loads.

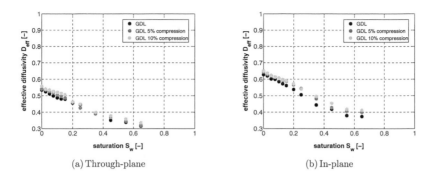

(a) Through-plane   (b) In-plane

Figure 3.30: *Calculated effective diffusivity of GDL with 10% PTFE under compression*

Experiments

In addition to the simulations, measurements of diffusivities for gas diffusion media were conducted. Several materials were examined (cf. appendix on page 113 for details) and compared with each other. With the help of the above described measurement devices different levels of compression were applied in steps of 5 µm to the samples.

First, the same gas diffusion layer type with a varying amount of hydrophobic coating was taken to examine the influence of the PTFE amount. Especially in the through-plane direction no influence could be observed. In the in-plane direction the effective diffusivity of the untreated material was slightly bigger, which can be explained by the distribution of the PTFE inside the GDLs. The SEM pictures of these materials show that the PTFE can mostly be found at the intersections of the fibres and only a small portion covers the fibres themselves. Hence the effective cross-section which influences the diffusion through the gas diffusion media is almost independent of the PTFE amount. In the in-plane direction (cf. also SEM cross-section on page 116) connections of PTFE among the single fibre layers in x-y direction are established. At the same time increasing PTFE content causes a reduction of the void space, which leads to a decrease of the effective diffusivity (cf. figure 3.31(b)).

In a second step the influence of a micro porous layer was identified: the three materials SGL24AA (GDL without PTFE, 230 µm thick), SGL24BA (GDL with PTFE, 230 µm thick), and SGL24BC (GDL with PTFE and MPL, 290 µm thick) were measured in

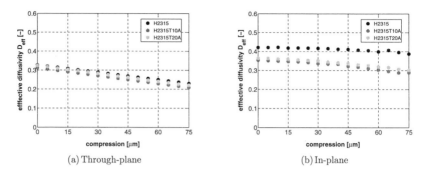

(a) Through-plane   (b) In-plane

Figure 3.31: *Comparison of measured effective diffusivities Freudenberg*

through- and in-plan direction. Again, only a small difference between the untreated SGL24AA and the SGL24BA could be observed. The additional PTFE causes a small decrease of the diffusivity. The curves of the GDL with MPL are significantly higher, especially in the through-plane direction, which is linked to the isotropic highly-porous micro porous layer with a thickness of ca. 60 µm on top of the macro porous fibrous GDL substrate. This layer for itself shows a significant higher diffusivity, which leads to an enhanced effective diffusivity of the complete material.

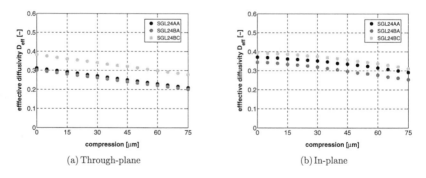

(a) Through-plane   (b) In-plane

Figure 3.32: *Comparison of measured effective diffusivities SGL Carbon*

Additional results for other materials can be found in the appendix on page 146 (pictures D.6 and D.7). The results of the measurements agree well with values from the literature [33, 58], which were partly obtained with other measurement strategies showing that the presented technique is able to capture the occurring diffusion process.

In the following chapter the measured and modelled constitutive relationships and transport parameters of PEMFC gas diffusion media are incorporated into a multi-component, two-phase Darcy model. The results are compared with experiments mimicking the counter-current flow of gas and water in thin hydrophobic porous layers, which arise at the cathode side of a polymer electrolyte membrane fuel cell.

4

Counter-current Flow in Thin Porous Layers

Counter-current flow of the gaseous and the liquid phase in porous gas diffusion media is the crucial point for high performance of polymer electrolyte membrane fuel cells. As discussed in the first chapter 1, the balance between sufficient water content in the membrane for satisfactory proton conductivity and preferably low saturation for optimal gas flow in the diffusion layers is the key for high power output of the fuel cell. As shown chapters 2 and 3 the water distribution and the resulting transport parameters and constitutive relationships of the GDLs have a strong influence on the transport processes. The present chapter will show an integral experiment and its simulation using previously determined relationships. A comparison of the experimental results with simulation data will round it off.

4.1 Counter-current Flow Experiment

The overall goal of the present chapter is the development and application of an experiment, which is able to mimic the counter-current flow situation of the liquid and gaseous phase mainly present in the cathode, especially inside the gas diffusion medium, of a polymer electrolyte membrane fuel cell. This special transport phenomenon stems from the water-forming overall reaction of the PEM fuel cell. In the first chapter of this thesis the basics of such fuel cells are sketched out and explained in detail. Due to the requirements of the counter-current flow experiment, several experiments were done in advance before the final concept and design was fixed. In the following the concept, the setup including utilised components and operation strategy, as well as a sketch of the reactor are given.

Figure 4.1 shows a cross-section of the developed reactor concept for the simulation of the counter-current flow. A channel with the width of 30 mm and the depth of 6 mm is milled on the upper side of a copper plate (thickness 15 mm, cf. also top view in figure 4.2). With the help of a PMMA plate a transparent cover is fixed on top. On the lower side of the copper plate there are cooling fins for the thermostatization of the plate. Again, another covering plate forms a channel which is flown-through with water. All the components are sealed against each other by O-ring sealings. The reacting species

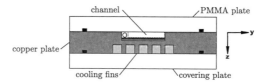

Figure 4.1: *Cross-section through reactor*

of oxygen, hydrogen, and nitrogen as inert gas are fed to the reactor, which is indicated on the middle axis in figure 4.2 as inlet and outlet at the beginning and the end of the channel. O_2 and H_2 diffuse through the porous layer to the surface of a catalytic active metal foil (light-grey layer in figure 4.3) and form water. The active metal foil has a platinum content in the range of $150 \, ^g/_{ft^3}$. The metal foil is bonded to the copper reactor with thermally conductive paste. On top of the catalytic layer a gas diffusion medium is placed (dark-grey layer). Identically to the fuel cell (cf. equations on page 2) water as reaction product is formed, at simultaneously a counter-current transport process of water and gas inside and through porous (mixed-wettable) media is established. The resulting heat release is handled via the cooling water flown-through channel on the lower side of the copper plate. At the beginning of the design phase simulations with COMSOL Multiphysicis were performed to optimize the design, so now almost isothermal conditions inside the complete reactor are given. Moreover, temperatures along the axial direction were measured and the largest difference of temperatures observed was 0.8 K. At the left side of the channel a tube (diameter of 3 mm) with a

Figure 4.2: *Top view of reactor*

small radial drilling (1.5 mm, cf. also figure 4.1, 4.2, and 4.3) forming a movable lance is installed. With the help of this additional tubing an axial profile of the gas concentrations can be measured. Again, COMSOL Multiphysics simulations in advance have shown that a steep concentration front is maintained while flowing through the reactor and the lance for axial concentration measurements. If axial profiles are measured via the lance, the plug valve at the end of the reactor is closed so that the total gas flux flows through the drilling into the tubing. These procedures ensure a relatively fast and constant signal of the measurement device.

The initial concentration is measured approximately at 25 mm in front of the GDL (dark-grey in figure 4.2, position marked with 1). Afterwards the lance (positioned at the side of the channel) is moved by 50 mm in the axial direction to the first measurement point above the GDL (indicated with 2). A period of 60 s ensures a constant gas concentration which flows through the gas analyser (cf. also figure 4.4). Then the lance is moved again by 50 mm in axial the x-direction and again 60 s are taken as measurement interval. The procedure is repeated until in sum the eight measurement positions described above are completed. The positions were defined in advance and periodically measured. The procedure is repeated several times to obtain a time- and spatial-dependent concentration profile of the complete process in the reactor. The transient behaviour of the system at the beginning as well as the stationary state including gas concentrations are a good indicator for the validity of the reactor model described in the next section.

Moreover, the reactor was embedded in a LabView-controlled environment where precise control and measurements of process parameters are possible. All the components of the testing rig were controlled via a robust Gantner e.pac DL controller, connected via Ethernet with a standard PC. A couple of Gantner e.bloxx modules (2x A3-1, 2x A4-1TC, 2x D2-1) were attached to the main controller providing an easy to use and reliable measurement system. The amount of water at the inlet and at the outlet of

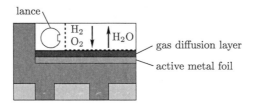

Figure 4.3: *Detailed cross-section*

the reactor and the moisture content respectively were measured with two Testo 6681 controller units with Testo 6613 humidity transmitters which are able to measure between 0 and 100 % relative humidity (accuracy 1 %, t_{90} max. 10 s) and are additionally equipped with a Pt1000 with a range of -40 up to 180 °C. The desired humidity level of the inflowing gas at the inlet was established via three MKS mass flow controllers (type 1179A, range of 0 up to 100 sccm for oxygen and hydrogen as well as 500 sccm for nitrogen), controlled by a MKS 647C controller unit. The correct portion of the feed gas flows through an electrically heated tube filled with catalytic pellets where a controlled water-forming oxy-hydrogen reaction takes place (cf. also figure 4.4 with the flow chart of the testing rig). The rest of the feed gas is also controlled via mass flow controllers with a maximum range of 500 sccm. A cryostat with water as well as

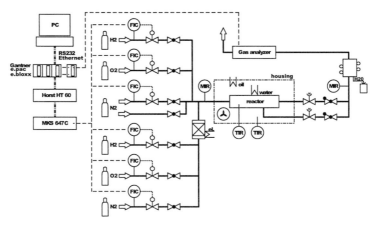

Figure 4.4: *P&ID of counter-current flow experiment*

an oil bath filled with silicone oil thermostats the reactor itself and the surrounding air in the housing to avoid unmeant condensation effects inside the reactor. An oil-flown twist of metal tubing as a heat exchanger is placed inside the housing where the air was continuously circulated with a small fan to establish the same thermal conditions all over the housing. In addition to the moisture levels, the gas concentrations were measured. Oxygen is consumed by the water-forming reaction, thus oxygen is appropriate as an indicator of the complete process. For the determination of the oxygen content a Servomex Xentra 4100 gas purity analyser was used, whose measurement principle is based on the paramagnetic properties of oxygen. The analyser has a measurement range of 0 up to 100 % and is pressure compensated. Moreover no sensibilities to other substances were found and the preciseness is in the range of 0.02 % with a linearity of 0.05 %. A repeatability of 0.01 % and a drift below 0.01 % per week ensured reliable measurement results over several periods of days. The response time was experimentally determined and is approximately 12 s at 200 sccm. Additionally the Servomex analyser was frequently re-calibrated with test gases to guarantee good quality of the results during the measurements.

4.2 Comparison with Darcy-approach

The measured data of the described experimental set-up were compared with a numerical model representing a multiphase-multicomponent Darcy-flow based approach. Here, the simulations were conducted with DuMu^X[27], a code based on the DUNE project [28]. The basis structure of the code was kindly provided by Andreas Lauser [61] within the framework of the international research training group NUPUS [72]. The necessary constitutive relationships and transport parameters were generally taken from the gained results described in the previous chapter.

The fundamental equations for the numerical model will be given in the following. A continuum approach on the macro-scale was chosen including the well-known REV assumption (cf. also chapter 1). The main application fields of these numerical models are oil and gas reservoir engineering, nuclear waste landfilling, and the modelling of hydrosystems including the propagation of pollutants [46]. Moreover, forecasting of scenarios or checks on measured values are within the scope of their possibilities.

But these large scale problems necessitate a scale-up from the micro-scale, where each phase (water, gas, solid) can be distinguished from the other ("fluid continuum" [11]), to the macro-scale where averaged quantities (e.g. saturation S_α of phase α or porosity ϵ) arise. With averaging and the resulting REV concepts the impracticality of detailed description of fluid flow through pores and throats of the porous medium is disposed of. Therefore balance equations on the micro-scale are integrated via an adequate averaging volume $\overline{V}$ ("REV") with a characteristic length l, (cf. figure 1.6 on page 14). The size of the REV must be as small as possible, but large enough to get significant statistical averages. *Bear* [11] as well as *Bear and Bachmat* [12] give a comprehensive overview of fundamentals about multiphase, multicomponent flow in porous media including averaging and arising quantities. Additionally, a summary of all the fundamentals for REV-based models is given in chapter 1.2. In the current chapter the validity of these averaged models for small-scale processes such as the counter-current flow of gas and water in thin (mixed-wettable) porous media (in the present case gas diffusion media of polymer electrolyte membrane fuel cells) is proven.

As described in the previous section oxygen, hydrogen, nitrogen, and water are fed to the reactor. The gaseous and liquid phases are considered separately. Hence a two-phase, four-component model in combination with continuum equations was set-up. All simulations are regarded as isothermal cases because the existing reactor concept ensures quasi-isothermal conditions. A pressure-saturation formulation of the problem was chosen. For more details regarding the formulation refer to *Helmig* [46].
In the present case the arising differential equations are solved on a 2D grid, which is regular and includes rectangular cells. A BOX scheme was applied to solve the arising differential equations. This scheme combines the advantages of a finite volume scheme (mass conservation) and finite element schemes (unstructured grids possible). In figure 4.5, the different areas of the underlying grid are shown. The flown-through channel on top was discretized with larger elements because in the lower part, where the porous medium is placed, the counter-current flow of gas and water takes place. At least four elements for the channel and eight or more elements in the y-direction for the porous section were applied to receive an appropriate resolution. The coupling of the gas flow through the channel and the porous layer may be done via the Beaver-Josephs criteria [13] but for the sake of simplicity both the gas diffusion medium at the bottom and the channel on top are treated as a porous medium. To achieve the same physical

behaviour as with the flow in a duct the permeability of the channel was calculated via the Hagen-Poiseuille equation. The transitions of liquid droplets out of the porous layer into the gas channel were neglected. Therefore small-scale simulations resolving these effects like SPH methods [49] are necessary.

In the x-direction (direction of flow) the number of elements was set based on the knowledge that quadratic elements guarantee the most precise results. On top and bottom of the simulation domain no-flow boundary conditions were applied. The complete edge on the left and right side are treated as Dirichlet boundaries. Intuitively one would apply a Neumann boundary condition at the outlet, but the early stage of the code framework did not allow the implementation of such a condition. As a kludge the domain was extended, i.e. the outlet part of the grid is to be at least twice as long as the inlet, ensuring a horizontal concentration curve progression past the active area, which is equivalent to a second-type boundary condition.

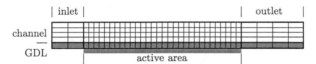

Figure 4.5: *Sketch of the chosen domain and grid*

For the active area (located between the inlet and outlet region) where the catalytic reaction takes place a simple Arrhenius approach was applied. On the lowest layer of the active area a reactive boundary layer was defined (cf. also light-grey line on the bottom of the pseudo-grid sketch in figure 4.5). The corresponding water source term q_r is expressed as the product of the reacting species multiplied by the Arrhenius rate constant $k(T)$.

$$q_r = c_{H_2} \cdot c_{O_2} \cdot k(T) \tag{4.1}$$

The parameters for the Arrhenius equation (cf. equation 4.2) were previously determined by simple water forming experiments at different temperature levels. Therefore the Pt-coated metal plate without any porous medium was overflown in the reactor with a diluted mixture of hydrogen and oxygen. In the following the resulting reaction rates were experimentally determined. Afterwards values were determined with the help of fitted curves.

$$k(T) = 1 \cdot 10^{-3} \cdot A \cdot e^{\frac{-E_A}{R \cdot T}} \tag{4.2}$$

The activation energy E_A is equal to $29\,739\,{}^{\text{J}}/_{\text{mol}}$, the prefactor A is $5.05 \times 10^{10}\,{}^{1}/_{\text{mol s}}$. The universal gas constant R has the value of $8.314\,{}^{\text{kJ}}/_{\text{mol K}}$.

Results

In the following the experimental and numerical results for gas diffusion media at different temperatures are compared. As described in the previous chapters, the amount and distribution of PTFE plays an important role for the transport of gas and water in GDLs as thin mixed-wettable porous media. As a typical gas diffusion medium a sample with 10 % PTFE was chosen. Therefore the corresponding constitutive relationships and parameters described in chapter 3 were implemented in the code. The analysis of the simulations was performed with ParaView and MATLAB. Subsequently water saturation profiles along the reactor axis are given. Moreover, the concentrations of oxygen as reacting species in x-direction are illustrated.

As given on page 103, the oxygen content inside the reactor was measured in the direction of the gas flow through the reactor during the experiments. Consequently, the profiles of the oxygen fractions are plotted vs. the axial direction of the reactor. Results from simulations and measurements are depicted together to show the validity of the chosen Darcy approach. For all simulations the constitutive relationships as well as parameters presented in the previous chapter (k_r-S_w, p_c-S_w, permeabilities, reaction rates etc.) are implemented in the above described Darcy-code. With respect to the hysteretic behaviour of the constitutive relationships an averaged value (i.e. centerline) of drainage and imbibition curves was chosen. The experimental results are marked with dots, the simulations are given as lines.

In figure 4.6 three measurements at different temperatures are compared with simulations. A gas diffusion medium with 10 % PTFE content was chosen as a key material. Only the active zone of the reactor where the water forming reaction takes place (between 0 m and 0.3 m) with two additional measurement points before and after the zone are illustrated. The inlet and outlet zones are not given.

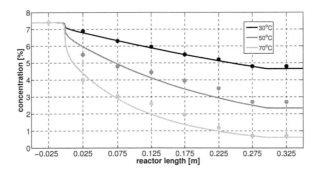

Figure 4.6: *Oxygen concentration at different temperatures*

The depicted concentration profiles came up within seconds after starting the experiment. No further change of the values and oxygen concentration profiles were observed during the measurements. The same behaviour could an be seen while running the above described Darcy code. And also with changed material properties reflecting non-hydrophobized GDLs as well GDLs with a higher PTFE content, there was an identical behaviour to be seen. Only the water saturation within the porous gas diffusion medium changed during the experiment. The global water content had roughly been checked during the preliminary experiments with a balance to get a figure about the saturation. As an outcome of the conducted simulations they will be also given in the subsequent paragraph.

As described above a hydrophobized GDL with 10 % PTFE content was implemented in the code. Therefore the capillary pressure-saturation relationship as well as the relative permeabilities based on the conducted measurements (cf. chapter 3) were implemented. Again, the axial direction is plotted on the x-axis whereas the water saturation is given on the y-axis. Figure 4.7 and 4.8 reflect the derived profiles. During the first minutes the saturation profile changed strongly, hence a detailed view is given. Every 30 s a profile was evaluated and plotted (cf. figure 4.7). After a short period a continuous increase of the saturation along the reactor axis can be seen. Steps of 4 minutes are given, the initial line represents the saturation profile after 2 minutes. An almost horizontally growing profile shows up and after approximately 60 minutes a stationary state is reached. Only a very tiny axial propagation of water inside the

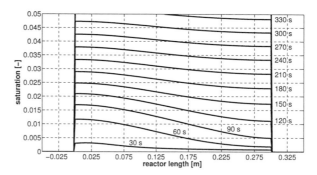

Figure 4.7: *Detailed saturation profiles*

porous zone into the non-reactive zones before and after the catalytic metal foil is monitored. Analogue shapes of curves were obtained with non-hydrophobized and hydrophobized GDLs (10 % PTFE content). Moreover stationary states were reached again after similar durations. Regarding the y-direction (which means perpendicular to the axial flow direction) an almost linear concentration gradient of oxygen and of the water saturation can be observed at all positions of the reactor. The experiments

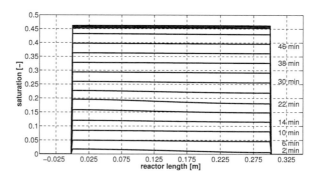

Figure 4.8: *Saturation profiles*

and the simulations based on a Darcy approach are in accordance with each other showing the validity of the experimental setup and the chosen approach. The constitutive relationships and transport parameters shown in the previous chapter have been included in the model. The fundamentals of the model including capillary-pressure saturation relationships, relative permeabilities, and effective diffusivities were sketched in the first chapter. As the water distribution inside the mixed-wettable porous gas diffusion layers is the most crucial point for the transport properties inside the GDLs, an optimization approach was chosen to simulate the wetting behaviour. Simulated annealing as a robust numerical scheme was selected to find the energetic minimum reflecting the water distribution inside virtually generated porous bodies under static conditions (cf. chapter 2). This data set was further processed via self-coded software and then utilized to calculate relative permeabilities of the gaseous phase as well as saturation-depending diffusivities. The experimental determined capillary-pressure saturation relationships and relative permeabilities of the water phase complete the set of information which is required to run the multiphase-multicomponent model of the porous medium. It has been also shown that the compression level of the gas diffusion layer plays an important role. Hence, all measurement devices are able to adjust the compression precisely in steps of microns. Moreover, multiple simulations were run to determine the above-mentioned transport parameters including the influence of compression. Based on these two approaches all necessary constitutive relationships and transport parameters of thin mixed-wettable porous media can be determined respectively measured. Moreover, it has been demonstrated that Darcy-flow based models for porous media are also applicable to thin technical porous layers.

As a drawback of this approach dynamic effects occurring during operation of a PEM fuel cell inside the gas diffusion layers are not considered. All simulated and measured values are gained under static conditions. To enlarge the understanding and knowledge furthermore, especially on the micro-scale, detailed information is needed about concentration profiles as well as the water contents in the different porous gas diffusion media during the experiment. Micro scale simulations, for example based on the smoothed particle hydrodynamics method, including mixed wettabilities will further enlarge the understanding for transport processes in technical porous media [48].
In the integral experiment this may be achieved via impedance spectroscopy, which is well-know in the field of fuel cell research. A multiple setup in sections tantamount to the reactor itself consisting of several sections independently connected to an impedance spectrometer and normally measured in parallel offers the possibility

of identifying axial profiles of the water saturation. Alternatively, nuclear magnetic resonance techniques could be used to identify the water distribution inside the porous GDL. Several researchers have applied cost-intensive techniques in related fields (for example water content and distribution of a PEM fuel cell under operation). Both setups will allow to localize and measure the water content and distribution inside the porous body on-line during the experiments. Additionally a visual observation of the processes by a camera system through a transparent cover from top of the setup would help to determine the position and the amount of emerging water droplets. The proposals will further enlarge the possibilities for comparisons with more detailed numerical models (e.g. models on the micro-scale) which help to increase the understanding of counter-current flow of gas and water in thin mixed-wettable porous media. Based on these findings progress in the optimization of transport process inside the gas diffusion media of polymer electrolyte membrane fuel cells will lead to a more efficient operation mode with a higher specific power output of PEMFC systems.

A
Gas Diffusion Media

The gas diffusion layer as an essential part of the polymer electrolyte membrane fuel cell can be characterized in several ways. Numerical and experimental ways of detailed analysis are presented in chapters 2 and 3. In the following sections first the applied techniques of experimental determination of material properties are described, then, on the basis fundamental parameters are given in tabular form in A.2. Additionally, two-dimensional (SEM) and 3D images (μ-CT) are presented.

A.1 Type of measuring methods

Subsequently the different techniques are shortly described and relevant information is given. With the exception of μ-CT imaging, all measurements were conducted at the University of Stuttgart.

μ-CT imaging

The μ-CT images were done with a SkyScan 1172 scanner manufactured by SkyScan (nowadays Bruker microCT N.V.). For details regarding material characterization with x-rays and CT imaging refer to *Lifshin* [64], and *Becker et al.* [14] especially for GDL. The x-ray source was operated with a source voltage of 40 kV and a source current of 250 μA. Each rotation step (0.25°) lasts 1.7 s leading to a scan duration of approximately 4 h. The scanner is equipped with a Hamamatsu C9300 11 MPx camera (2672 x 4000 pixel, 16 bits depth), which leads to the resulting voxel size of 0.73 μm.

The raw data is given as a sequence of jpeg pictures each representing a slice of the data set with a thickness of 0.73 μm. The jpeg pictures are converted to greyscale bitmaps which are processed by a proprietary MATLAB code building geometry representing files for simulations as well as vtk data for the visualization of the structures. The major task of the code is the correct setting of the grey-scale threshold value for the binarization. For this purpose different techniques are implemented: first, the correct threshold value is adjusted via the known porosity of the structure, second, the threshold value itself is set and third, Otsu's method [79] is applied as a widely-used technique for binarization where the the variance of background and foreground pixels is considered. Within each class of pixels the variances are minimized and concurrently the variance between both classes is maximized. Figure A.1 shows an almost complete sample with the dimensions of 1520 x 1520 x 300 voxels. Moreover,

Figure A.1: *Large extract of a μ-CT image SIGRACET GDL 24 BA*

the resulting structures can examined virtually, e.g. based on morphological operations (e.g. granulometry) [91] the pore size distribution inside the REV can be determined. For this the size of the pores is approximated by growing virtual spheres. If such a sphere collides with the wall of the pore, the diameter of the sphere is taken as pore diameter. These are the tools that are coded to verify and support the modelling of constitutive relationships and parameters for the examined gas diffusion media.

Material characterization

The fibre diameters and the spatial distribution of the hydrophobic PTFE were determined with the help of scanning electron microscope (SEM) pictures. The diameters of the fibres were measured on the diagonal of each picture taken from the different material samples. Moreover, different, material-depending distributions of PTFE were detected and documented. The thicknesses of the materials were measured with micrometer calliper: to this end the screw was closed until the sample stuck between both sides of the calliper. For SEM imaging the samples of the GDLs were fixed on a carrier; the cross-sections of the gas diffusion media were obtained by cutting the sample with a keen scalpel in advance.

Porometry

The porosity of each material sample was evaluated through decane porometry. Decane as a hydrocarbon wets almost all surfaces and has a low surface tension of approximately $23.8\,\mathrm{mN/m}$. Each sample was cut with a hollow punch and weighed in advance with an analytical balance. With the help of this information the area weight m_a can also easily be determined. The thicknesses of the materials were also known. The dry sample was then immersed in n-decane for a while and carefully shaken. The filled and wetted sample was removed from the liquid n-decane and directly weighed. For each material the procedure was repeated five times with different samples. With the information gained the increase of weight can be measured and evaluated to find the porosity of the sample under the assumption that all pores and cavities are filled with the hydrocarbon. The boiling point of n-decane is sufficiently high ($174\,°C$), therefore the error due to evaporation of decane during the measurement is small. Additionally, the resulting porosity of the binarized μ-CT images (with Otsu's method) was also taken into consideration to evaluate and verify the measurements with decane.

A.2 Material properties

In the following SEM pictures as well as some results of μ-CT imaging of several different gas diffusion media are presented. Fibre radii distributions, mean fibre diameters, material thicknesses, porosities, and area weights are given in tabular form on the next pages for each material.

H2315	(Freudenberg)

SEM image - top view (200x)	SEM image - cross-section

μ-CT image	detail of μ-CT image

Fibre radii distribution

Mean fibre diameter $\overline{d}$ with σ [µm]:	10.39 ± 0.56
Thickness d [µm]:	250
Porosity ϕ with σ [%]:	75.7 ± 2.89
Area weight m_a [g/m²]:	100.4

Table A.1: *Properties of Freudenberg H2315*

H2315T10A   (Freudenberg)

SEM image - top view (200x)   SEM image - cross-section

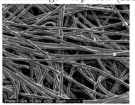   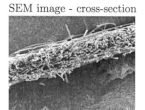

μ-CT image   detail of μ-CT image

Fibre radii distribution

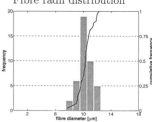

Mean fibre diameter $\overline{d}$ with σ [µm]:	10.27 ± 0.98
Thickness d [µm]:	230
Porosity ϕ with σ [%]:	70.3 ± 2.10
Area weight m_a [g/m²]:	107.2

Table A.2: *Properties of Freudenberg H2315T10A*

H2315T20A   (Freudenberg)

SEM image - top view (200x)

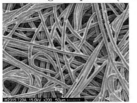

SEM image - cross-section

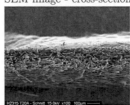

μ-CT image

detail of μ-CT image

Fibre radii distribution

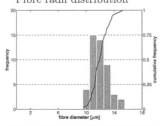

Mean fibre diameter $\overline{d}$ with σ [µm]:	12.00 ± 1.18
Thickness d [µm]:	240
Porosity ϕ with σ [%]:	63.9 ± 1.31
Area weight m_a [g/m²]:	120.8

Table A.3: *Properties of Freudenberg H2315T20A*

H2315IX53   (Freudenberg)

SEM image - top view (200x)

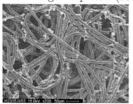

SEM image - cross-section

μ-CT image

detail of μ-CT image

Fibre radii distribution

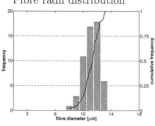

Mean fibre diameter $\overline{d}$ with σ [µm]:	11.20 ± 1.13
Thickness d [µm]:	240
Porosity ϕ with σ [%]:	66.8 ± 1.35
Area weight m_a [g/m²]:	104.5

Table A.4: *Properties of Freudenberg H2315IX53*

H2315T10AC1   (Freudenberg)

SEM image - top view (200x)   SEM image - cross-section

   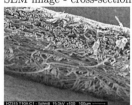

μ-CT image   detail of μ-CT image

Fibre radii distribution

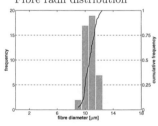

Mean fibre diameter $\overline{d}$ with σ [µm]:	10.68 ± 0.81
Thickness d [µm]:	300
Porosity ϕ with σ [%]:	68.1 ± 0.98
Area weight m_a [g/m²]:	149.9

Table A.5: *Properties of Freudenberg H2315T10AC1*

SIGRACET GDL 24AA	(SGL Carbon)

SEM image - top view (200x)

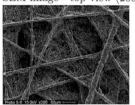

SEM image - cross-section

$\rightarrow$ due to the same substrate, a similar cross-section to SGL 24BA (with PTFE) was observed, cf. also next page.

μ-CT image

detail of μ-CT image

Fibre radii distribution

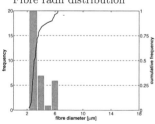

Mean fibre diameter $\overline{d}$ with σ [µm]:	3.47 ± 1.06
Thickness d [µm]:	230
Area weight m_a [g/m²]:	58.3

Table A.6: *Properties of SGL Carbon SIGRACET 24AA*

SIGRACET GDL 24BA (SGL Carbon)

SEM image - top view (200x)

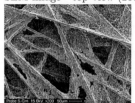

SEM image - cross-section

μ-CT image

detail of μ-CT image

Fibre radii distribution

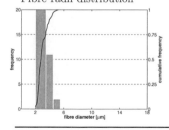

Mean fibre diameter $\overline{d}$ with σ [µm]:	3.00 ± 0.71
Thickness d [µm]:	230
Area weight m_a [g/m²]:	54.9

Table A.7: *Properties of SGL Carbon SIGRACET 24BA*

SIGRACET GDL 24BC	(SGL Carbon)

SEM image - top view (200x)	SEM image - cross-section
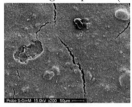	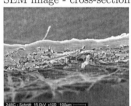
μ-CT image	detail of μ-CT image

Fibre radii distribution

$\rightarrow$ due to the same substrate, a similar fibre radii distribution to SGL 24BA was observed, cf. also previous page.

Mean fibre diameter $\overline{d}$ with σ [μm]:	cf. previous page, Sigracet GDL 24 BA
Thickness d [μm]:	290
Area weight m_a [g/m²]:	103.1

Table A.8: *Properties of SGL Carbon SIGRACET 24BC*

B
Algorithms for Structure Generation

Modelling and simulation of the water distribution inside porous, mixed-wettable layers requires algorithms which are capable to generate virtual porous materials owning properties close to reality. The virtual structure generation of non-woven (B.1) as well as woven gas diffusion layers (B.2) is described in this chapter. Moreover different algorithms for the insertion of PTFE as non-wetting coating, both located on the surface of fibres and in regions of intersection of fibres, are presented and discussed (cf. subsection B.3). Finally, possible sources of error during the discretization of generated porous media are critically examined and evaluated.

All the algorithms for the virtual material design were implemented in C to ensure good performance; the visualization is based on VTK standard [95] and conducted with the help of ParaView [80]. The desired structure properties are handed over by parameter files (cf. listings B.2 and B.3). The different types of coatings and output formats are controlled via parameters on the command line of a linux system. Moreover structural information is stored in a separate file and can be reloaded, e.g. to generate the same structure with a different spatial resolution or a different technique for adding PTFE. In addition, an algorithm imitating the compression of the generated diffusion media, was developed and implemented. In the following a brief description of it is given: assuming a compression in the z-direction, first all layers in the xy-plane filled only with gas are removed. If the removal of gaseous xy-planes is not sufficient, a second step is conducted. Randomly chosen, equally distributed gas elements are removed until the desired compression level is achieved. The solid and PTFE elements slide closer to each other ("tetris principle"); hence a small deformation of the structure under compression load has to be accepted.

File format

In the following a short description of the proprietary file format which was used for the representation of virtually generated materials and interfacial areas is given. The main idea behind the format is a structure enabling quick modifications during testing phases as well as easy adaption of new code. To fulfil these criteria an ASCII format is chosen. Listing B.1 gives a small example. On page 54ff in chapter 3 the diverse usages of the file format are illustrated.

Listing B.1: *Proprietary file format*

```
 1   ─────────────────────────────────────────
 2   general structure:
 3   ─────────────────────────────────────────
 4   [x y z]
 5   x_1, y_1, z_1 ...          x_i, y_1, z_1
 6   ...
 7   x_1, y_j, z_1 ...          x_i, y_j, z_1
 8
 9   ...
10
11   x_1, y_1, z_k ...          x_i, y_1, z_k
12   ...
13   x_1, y_j, z_k ...          x_i, y_j, z_k
14
15
16   ─────────────────────────────────────────
17   small example:
18   ─────────────────────────────────────────
19   4         5         3
20   1         1         1         1
21   0         0         0         3
22   0         2         2         3
23   0         2         2         2
24   1         1         1         1
25   0         0         0         0
```

26	1	1	1	1
27	2	0	0	3
28	2	0	0	0
29	0	0	0	1
30	0	1	1	1
31	0	0	0	0
32	1	2	2	2
33	1	2	2	2
34	1	1	1	1

```
35
36  with :
37  0 ==> solid           1 ==> gas
38  2 ==> liquid          3 ==> PTFE
```

B.1 Non-woven structures / carbon paper

The non-woven structures, also referred to as carbon papers, are porous media without regular repeating characteristics. Single fibres lie on each other and form a highly porous material with anisotropic properties (cf. also chapter A for more details). In the following the kind of description, the underlying mathematics and the chosen parameters are presented.

The non-woven structure with the domain size, the resolution, the preferred direction and radius of fibres as well as the corresponding mean variations and the resulting porosity can be chosen via parameter file (cf. listing B.2). The addition of hydrophobic coating is also regulated with the help of command line parameters and the last six lines of the parameter file: all the combinations of the three methods of PTFE generation (cf. section B.3) are possible. The distance between fibres for coating and the width of a parabola are used by the algorithm generating parabolic connections (B.3.1). The next three lines in the file are applied to the stochastic generation of hydrophobic surface patches (B.3.2), while the last one represents the decisive radius for the morphological operation (B.3.3).

The first step of virtual GDL generation (*genGDL*) is the estimation of numbers of fibres for the precise attainment of the desired porosity. The next step is the creation of Gaussian distributed fibre radii to accommodate the natural deviation during the fabrication. Then, an uniformly distributed starting point inside the considered modelling domain is stochastically chosen. Based on the given parameters and standard deviations, the direction of each fibre is generated with the help of normal distribution. The carbon fibres are modelled as straight cylinders (mathematically represented by their centre lines) with a constant diameter. They are extended and afterwards cut at the lateral faces of the domain. Finally the continuous description of the fibres is terminated and transformed into a discrete representation on a grid with predefined spacing. The resulting representation is stored in an ascii file. To this end a specific number is assigned to each material: solid = 0; gas = 1; liquid = 2; PTFE = 3. In the first line of the file the dimensions of the structure are specified; the three-dimensional entity itself is stored in cross-sections of the x-y-plane (cf. also listing B.1).

Listing B.2: *Parameter file for non-woven structures*

```
1    0                    ! – 0: use parameter file , 1: do reimport
2    75.0                 ! – length of REV in x–direction [µm] (double)
3    75.0                 ! – length of REV in y–direction [µm] (double)
4    200.0                ! – length of REV in z–direction [µm] (double)
5    75                   ! – resolution in x–direction [voxel] (int)
6    75                   ! – resolution in x–direction [voxel] (int)
7    200                  ! – resolution in x–direction [voxel] (int)
8    0.250                ! – preferred direction x (double)
9    0.250                ! – preferred direction y (double)
10   0.0                  ! – preferred direction z (double)
11   0.75                 ! – standard deviation of preferred direction x (double)
12   0.75                 ! – standard deviation of preferred direction y (double)
13   0.1                  ! – standard deviation of preferred direction z (double)
14   78.0                 ! – porosity [%] (double)
15   3.5                  ! – mean fibre radius [µm] (double)
16   1.0                  ! – standard deviation of mean radius [µm] (double)
17   0.25                 ! – distance between fibres for coating (double)
18   8.0                  ! – width of parabola for coating (double)
19   35.0                 ! – surface coverage of coating [%] (double)
20   250.0                ! – number of elements per coating patch (double)
21   3.0                  ! – standard deviation of elements per patch (double)
22   5.0                  ! – radius for closing algorithm [voxel] (double)
```

B.2 Woven structures / carbon cloth

A second class of fibrous porous media are the woven structures, also labelled as carbon cloth. This subsection shows how they can be generated. The parameter file (cf.
listing B.3) with the length and resolution of the domain, diameter and preferred direction as well as associated standard deviation of the fibres, size and properties of fibre
bundles, and desired characteristics of the hydrophobic coating is given below. The
parameter file together with command line entries make it possible to generate woven
structures (*genCDL*) with exactly defined properties. The hydrophobic coating can be
generated via morphological closing (pictured in the next subsection) or stochastically
distributed PTFE patches on the surface.

Initially the quantity of fibres inside one bundle is estimated via the volume of the
single fibres, the chosen porosity, and the number of bundles as well as their spatial
extent. The variation of orientation is similar to the non-wovens modelled by standard
deviations in combination with given preferred directions. The next step is to generate
a normal distribution with the estimated number of fibres based on the presetting.
Subsequently the starting points of fibres on the lateral face of a fibre bundle are generated with the help of a random number generator producing uniformly distributed
numbers. The woven fibre mat with its perpendicular bundles lies in the x-y-plane.
The fibres (mathematically represented by their centre lines) inside the first bundle
parallel to the y-axis are generated with the help of a trigonometric formulation. The
position of the fibre centre line in z-direction along the y-axis is modelled with:

$$z_x = A \cdot sin(B \cdot y + C) \tag{B.1}$$

If the generated fibre lies inside the virtual box surrounding the bundle, the fibre is
discretized and stored in a matrix. The described procedure is repeated until the desired bundle porosity is achieved. The parameters A, B, and C are computed with
respect to the number of bundles and their geometrical size given in the parameter
file. To generate an alternating woven structure with the requested number of bundles,
the complete fibre bundle (parallel to the y-axis) is copied and moved to new positions.

The fibre bundles along the y-axis are created in the same manner - each fibre centre line is represented by a sine wave:

$$z_y = A \cdot sin(B \cdot x + C) \tag{B.2}$$

Again, the condition of belonging to the fibre bundle is checked and in case of pass the fibre is discretized and stored in the global matrix. Exactly as with the y-direction, this approach is repeated until the bundle porosity is reached, and finally the whole bundle is copied and moved to the other positions parallel to the original one.

Listing B.3: *Parameter file for woven structures*

```
1   0          ! - 0: use parameter file , 1: do reimport
2   200.0      ! - length of REV in x-direction [µm] (double)
3   200.0      ! - length of REV in y-direction [µm] (double)
4   40.0       ! - length of REV in z-direction [µm] (double)
5   200        ! - resolution in x-direction [voxel] (int)
6   200        ! - resolution in y-direction [voxel] (int)
7   40         ! - resolution in z-direction [voxel] (int)
8   0.0        ! - preferred direction x (double)
9   1.0        ! - preferred direction y (double)
10  0.0        ! - preferred direction z (double)
11  0.0        ! - standard deviation of preferred direction x (double)
12  0.2        ! - standard deviation of preferred direction y (double)
13  0.0        ! - standard deviation of preferred direction z (double)
14  70.0       ! - porosity [%] (double)
15  4.4        ! - mean fibre radius [µm] (double)
16  0.5        ! - standard deviation of mean radius [µm] (double)
17  30         ! - width of fibre bundle [µm] (double)
18  30         ! - heigth of fibre bundle [µm] (double)
19  75.0       ! - density of the fibre bundle [%] (double)
20  6          ! - number of bundles in each direction (int)
21  5.0        ! - radius for closing algorithm [voxel] (double)
22  20         ! - surface coverage of coating [%] (double)
23  100        ! - number of elements per coating patch (double)
24  3.0        ! - standard deviation of elements per patch (double)
```

In figure B.1 an example of a virtually generated woven structure coated with PTFE is given. The parameters are identical to those in listing B.3.

Figure B.1: *Example of generated woven structure with PTFE*

B.3 Hydrophobic coating

In the previous sections the generation of woven and non-woven structures is explained in detail. In reality the carbon fibres with a contact angle of 86° with water are treated with a hydrophobic substance to enhance the water removal inside the porous backings of polymer electrolyte membrane fuel cells. Mostly, polytetrafluoroethylene is applied to the fibrous framework to increase the contact angle and the repellency against water. Several ways of coating are known, the most common ones used are the dipping or spraying of an aqueous solution with PTFE particles combined with a subsequent heat treatment to fix the PTFE particles. At the beginning of this chapter different scanning electron microscopy (SEM) pictures show the spatial distribution of PTFE allover the fibres. Most of the coating can be found at the intersections of fibres (cf. figure B.2(a)) where it forms parabolic connections. The rest of the hydrophobic coating is located on the surface of the fibres (figure B.2(b)). The following subsections present different approaches to the application of PTFE inside gas diffusion media with the goal of realistic structure generation.

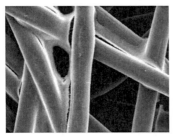

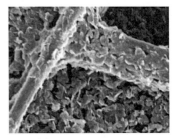

(a) PTFE parabolic between fibres   (b) PTFE on fibre surface

Figure B.2: *Examples of PTFE distribution*

B.3.1 Parabolic between fibres

As mentioned above and pictured in figure B.2(a), the main part of PTFE can be found
at the intersections of fibres in the shape of parabolas. The proper way to imitate them
is described in this subsection. Each fibre is generally represented by its centre line
(dotted lines in figures B.3, (a)-(c)). The algorithm first searches for intersections of
fibres. Skew fibres (neighbouring, but not intersecting) can also be detected - therefore
the scanned distance between fibres can be varied with the help of a parameter given in
the file. When an intersection between two fibres comes up, the normal vector for the
determination of the shortest distance is identified. Figure B.3 shows only two fibres
intersecting each other directly to clarify the subsequent explanation. The shortest
distance in this case is equal to zero. Secondly, the bisecting line between the centre
lines of the two fibres is computed. After that the vertical straight line through the
vertex V of the sought parabola and the midpoint of the connecting normal vector
is determined (cf. line l in figure (b)). The vertices (C_{top} and C_{bottom}, figure (b)) on
each fibre are also determined (points; the connections between them and intersection
point of the parabola with the fibre (cf. points A_1 and A_2 in figure (a)) represent the
outline of the inserted parabolas. Together they form the envelope for the hydrophobic
material. The last step in the algorithm is to fill the area between the envelope and the
two fibres with the hydrophobic substance, which is shown in figure (c). As an example
the resulting configuration is shown in figure B.4 where two fibres are intersecting. The
hydrophobic coating can be found as a parabola between the edges of the fibres.

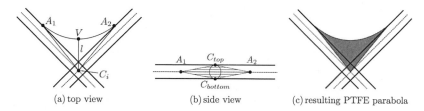

(a) top view   (b) side view   (c) resulting PTFE parabola

Figure B.3: *Example of PTFE generation with parabolas*

Figure B.4: *Parabolic PTFE between intersecting fibres*

B.3.2 Stochastically distributed

In order to account for the polytetrafluoroethylene located on the surface of the fibres (cf. figure B.2(b)), a second approach is introduced. It works with non-woven as well as with woven fibrous porous media. Stochastically distributed patches with a certain size are generated via a simple algorithm described in the following. The size of the hydrophobic patches is given by the diameter specified in the parameter file. The amount of PTFE on the surface is also controlled by the file: the ratio of covered fibre surface to the whole surface can be chosen.

First the surface of all fibres has to be determined and stored. Uniformly distributed random numbers in the interval of $[0 \ max_i]$, where max_i is the last voxel in each dimension of the domain, are employed to choose a surface element which acts as the centre $M(x, y, z)$ of the patch. Subsequently points $P_i(x, y, z)$ are randomly chosen in the same interval $[0 \ max_i]$. The resulting distances $\Delta x, \Delta y, \Delta z$ between M and P_i in each spatial direction are inserted into a Boltzmann distribution with a random number r and compared to the decisive radius r.

$$r < e^{-\sqrt{\Delta x^2 + \Delta y^2 + \Delta z^2} - \sqrt{3}} \tag{B.3}$$

If r is smaller than the right-hand side, the surface point is accepted and taken over. The algorithm is repeated until the desired size and amount of patches is reached. Figure B.5 gives an impression of such stochastically distributed patches of PTFE.

Figure B.5: *Stochastically distributed PTFE at intersecting fibres*

B.3.3 Morphological closing

The hydrophobic connections between fibres and their interspaces can also be created by morphological operations. The technique of mathematical morphology (MM) was invented by Georges Matheron and Jean Serra in 1964 and was further developed and expanded till today. *Serra* [86] and *Soille* [91] give both a good overview about fundamentals and applications of mathematical morphology.

In the following the chosen approach [48] for the generation of hydrophobic parts of the mixed-wettable structure is sketched. The main idea is the usage of discrete, three-dimensional data-sets based on voxels with different properties. For this a closing operation is applied to generate fibre conglomerates. Morphological closing is defined as a sequence of dilation followed by an erosion.

The dilation δ of an image Q by a structuring element e as one basic operation of MM for a locus of points $\mathbf{x}$ is given by:

$$\delta_e(Q) = Q \oplus e = \{\mathbf{x} \mid e_x \subseteq Q\} \tag{B.4}$$

Erosion ϵ of an image Q by a structuring element e as another basic operation, with respect to a locus of points $\mathbf{x}$ is defined as:

$$\epsilon_e(Q) = Q \ominus e = \{\mathbf{x} \mid e_x \cap Q \neq \emptyset\} \tag{B.5}$$

Figure B.6 depicts the two steps: in subfigure (a) the dilation operation for two fibres (dark squares) with different radii in 2D is sketched. A structuring element with a diameter of two voxels is applied to the surface of the fibres. Therefore, the surface of fibres has to be detected in advance. New elements show up (cf. the squares with waves in (a)) and will be stored temporarily. Subsequently an erosion with the same diameter of structuring element is applied to the resulting interface of the temporal structure (cf. subfigure (b)). For this the interface between the squares with waves and the gaseous phase (white squares) must be detected. Some elements remain after the erosion operation and and will be finally switched to coating elements (light grey squares in figure B.6(b)). A simple 2D example of the described algorithm is given in figure B.6 on page 136: two fibres are close to each other and an additional PTFE phase is built between the edges of the fibres.

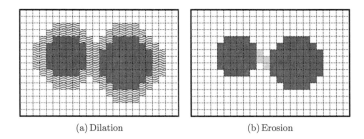

(a) Dilation   (b) Erosion

Figure B.6: *Closing between two fibres*

Figure B.7: *Morphological closing at intersecting fibres*

B.4 Error analysis

The algorithms presented in B.1 and B.2 generate discrete, voxel-based volume data files. The generation of structures itself however is done in a continuous mode, i.e. the whole description during the generation process is done by analytical functions and vectors. While transferring the steady functions to discontinuous geometry files which are based on cubic elements, a systematic error occurs [51]. It is apparently due to the limited resolution of the discrete representation. In the following the deviation of the discrete from the analytical, continuous form is given.

B.4.1 2D analysis

In the beginning, by virtue of complexity, the three-dimensional fibres are represented by two-dimensional cross-sections. An ensemble of different diameters is studied and the deviations are given in table B.1. The relative error ϵ is derived from the comparison of the value of the discretized fibre cross-section A with the analytical result for a cylinder with diameter $d = 2r$ in continuous space.

B.4.2 3D analysis

Extending the 2D analysis to three dimensions is straightforward: again a single fibre, generated in continuous space is discretized and evaluated with respect to the analytically derived volume. Therefore a stochastic approach is chosen, which is described in the following section. Moreover, different sizes of voxels χ starting with $0.25\,\mu m$ up to $2\,\mu m$ are chosen, to evaluate the influence of the voxel size / discretization.

2D representation	d [µm]	A [µm^2]	ϵ [%]
	5	19.6	-22.7
	6	28.3	11.6
	8	50.3	3.3
	10	78.5	1.8
	12	113.1	-1.0

Table B.1: *2D error analysis of different discretized fibre diameters*

Single fibres

Knowing the error during discretization of the generated structures is one key factor to estimate the global error of virtual material design. In case of permanent under- or overestimation for a single fibre, the number of fibres in a REV compared to real structures will consequently be predicted inaccurately. Therefore the generation of a single fibre is repeated 100 000 times and the resulting discretized volume is compared with the analytical derived one for each fibre. The domain for the investigation takes 50 x 50 x 50 voxels. Mean error μ_ϵ and its standard deviation σ_ϵ are given in table B.2 for different voxel sizes. The resulting values, especially the standard deviation of the error, show that with increasing voxel size χ the impreciseness of discretization grows as expected. Though, the resulting error is acceptable for all kinds of fibre diameters under realistic conditions. These errors are taken into account with the help of the utilized parameter file.

d [µm]	χ [µm]	μ_ϵ [%]	σ_ϵ [%]
5	0.25	2.20	2.31
	0.5	−0.15	4.13
	1.0	5.54	5.24
	1.5	6.27	3.48
	2.0	7.93	4.84
6	0.25	1.36	2.11
	0.5	2.81	2.82
	1.0	−3.83	2.94
	1.5	−4.18	3.69
	2.0	2.00	8.53
8	0.25	1.59	1.77
	0.5	1.74	2.53
	1.0	−0.49	4.04
	1.5	3.45	5.06
	2.0	−5.35	5.47
10	0.25	1.28	1.67
	0.5	1.99	2.36
	1.0	−0.34	4.16
	1.5	7.31	2.63
	2.0	8.98	6.18
12	0.25	0.78	1.78
	0.5	1.07	2.21
	1.0	2.63	2.82
	1.5	−1.24	3.82
	2.0	−4.52	3.97

Table B.2: *3D error analysis of different discretized fibre diameters*

C

Parameters and Values

C.1 k_r-S_w measurement

In chapter 3 (pages 53ff) the measurement as well as the numerical determination of k_r-S_w relationships are described. The correct choice of the flow regime, depending on the dimensionless capillary number Ca and on Reynolds number Re (defined in chapter 1 on pages 16 and 21), are shown in the following. Capillary number Ca has to be below 1×10^{-6} in case of fuel cells, due to the small velocity v of the water phase inside the porous medium which is also known as capillary fingering. Velocity v is expressed as:

$$v = \frac{Ca \cdot \sigma}{\eta} \tag{C.1}$$

Because of the anisotropic properties of gas diffusion layers, in- and through-plane measurements were done. The shape of the samples is given in chapter 3 on page 75, figure 3.13. For the through-plane measurements the sample diameter is equal 13.7 mm; for the in-plane measurement the inner diameter has 13.6 mm and the outer one 40.0 mm.

In the following an estimation of maximum water fluxes is given while ensuring the correct flow regime. As the lower boundary the capillary number $Ca_{min} = 1 \times 10^{-8}$ was chosen, the upper one is equal to $Ca_{max} = 1 \times 10^{-6}$. With the help of the flown-through area A and the viscosity of water $\eta_w^{25°C} = 0.891 \times 10^{-3}$ Pa s the flux Q of water with respect to Ca_{min} and Ca_{max} can be determined. These values were taken as limits for the experimental determination of relative permeabilities. All chosen fluxes lay between these constraints:

- **through-plane:**
  $Q_{min} = 7.22 \times 10^{-3}$ ml/min
  $Q_{max} = 0.722$ ml/min

- **in-plane:**
  $Q_{min} = 8.26 \times 10^{-4}$ ml/min
  $Q_{max} = 0.0826$ ml/min

For these small volume fluxes the corresponding Reynolds number Re in both cases is smaller than 1, hence the application of the results from the experimental determination of k_r-S_w relationships with Darcy's law is allowed [26].

C.2 D_{eff} measurement

For the calculation of the effective diffusivity the bulk diffusion coefficient is needed. In the present case the empirical correlation proposed by *Fuller et al.* [34], [35], [36] was used. Based on the atomic diffusion volumes almost all diffusion coefficients for binary gas systems at low pressures can be derived.

$$D_{ij} = \frac{0.00143 \cdot T^{1.75}}{p \cdot M_{ij}^{\frac{1}{2}} \cdot \left[(\Sigma_\nu)_i^{\frac{1}{3}} + (\Sigma_\nu)_j^{\frac{1}{3}} \right]^2} \tag{C.2}$$

where:

$$
\begin{aligned}
D_{ij} &= \text{binary diffusion coefficient } [\text{cm}^2/\text{s}] \\
T &= \text{temperature [K]} \\
M_i, M_j &= \text{molecular weights of species } i \text{ and } j \text{ [g/mol]} \\
p &= \text{pressure [bar]} \\
M_{ij} &= 2 \cdot \left[\left(\frac{1}{M_i} \right) + \left(\frac{1}{M_j} \right) \right]^{-1} \text{ [g/mol]}
\end{aligned}
$$

D
Additional results

In the following results of supplemental measurements for different types of gas diffusion media are presented. First, the preciseness of the capillary pressure–saturation relationship measurements is demonstrated based on a material (GDL with MPL and 10 % PTFE, SGL24BC) those three different samples has been measured and plotted together. Second, additional permeabilities of gas diffusion media for various materials under compression for both principal directions are given. Third, effective diffusivities of GDLs in through- and in-plane direction are shown.

D.1 Capillary pressure–saturation relationship

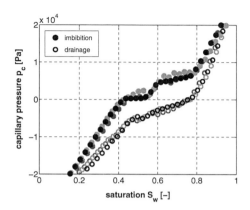

Figure D.1: *Repeatability of p_c-S_w measurement*

D.2 Permeability

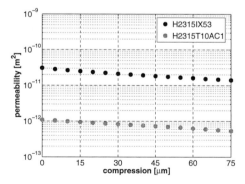

Figure D.2: *Comparison of through-plane permeabilities (Freudenberg)*

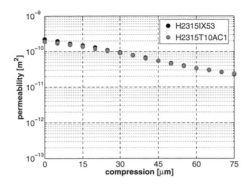

Figure D.3: *Comparison of in-plane permeabilities (Freudenberg)*

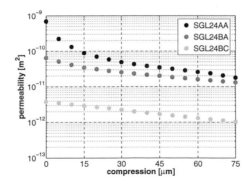

Figure D.4: *Comparison of through-plane permeabilities (SGL Carbon)*

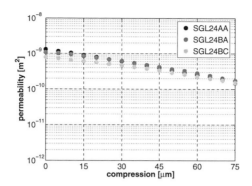

Figure D.5: *Comparison of in-plane permeabilities (SGL Carbon)*

D.3 (Effective) Diffusivity

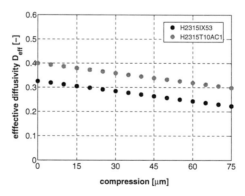

Figure D.6: *Comparison of through-plane diffusivities (Freudenberg)*

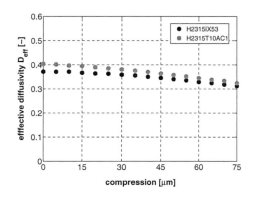

Figure D.7: *Comparison of in-plane diffusivities (Freudenberg)*

Bibliography

[1] AARTS, E., DE BONT, F., HABERS, J. and VAN LAARHOVEN, P. A parallel statistical cooling algorithm. *Lecture Notes in Computer Science*, 210: 87–97, **1986**.

[2] AARTS, E. and KORST, J. Simulated Annealing and Boltzmann Machines (John Wiley & Sons), **1989**.

[3] ACOSTA, M., MERTEN, C., EIGENBERGER, G., CLASS, H., HELMIG, R., THOBEN, B. and MÜLLER-STEINHAGEN, H. Modeling non-isothermal two-phase multicomponent flow in the cathode of PEM fuel cells. *Journal of Power Sources*, 159: 1123–1141, **2006**.

[4] ADAMSON, A. W. Physical Chemistry of Surfaces (Wiley-Interscience), **1990**.

[5] ALLWRIGHT, J. R. A. and CARPENTER, D. B. A distributed implementation of simulated annealing for the travelling salesman problem. *Parallel Computing*, 10(3): 335 – 338, **1989**.

[6] ASTM INTERNATIONAL. D2434: Standard Test Method for Permeability of Granular Soils (Constant Head). **2006**.

[7] ASTM INTERNATIONAL. D5856: Standard Test Method for Measurement of Hydraulic Conductivity of Porous Material Using a Rigid-Wall, Compaction-Mold Permeameter. **2007**.

149

[8] ASTM INTERNATIONAL. D5084: Standard Test Methods for Measurement of Hydraulic Conductivity of Saturated Porous Materials Using a Flexible Wall Permeameter. **2010**.

[9] AZENCOTT, R. Simulated annealing: Parallelization techniques (Wiley), **1992**.

[10] BAZYLAK, A., SINTON, D., LIU, Z.-S. and DJILALI, N. Effect of compression on liquid water transport and microstructure of PEMFC gas diffusion layers. *Journal of Power Sources*, 163: 784–792, **2007**.

[11] BEAR, J. Dynamics of fluids in porous media (Elsevier, New York), **1972**.

[12] BEAR, J. and BACHMAT, Y. Introduction to Modeling of Transport Phenomena in Porous Media (Kluwer Academic Publishers), **1990**.

[13] BEAVERS, G. S. and JOSEPH, D. D. Boundary conditions at a naturally permeable wall. *Journal of Fluid Mechanics*, 30: 197–207, **1967**.

[14] BECKER, J., SCHULZ, V. and WIEGMANN, A. Numerical Determination of Two-Phase Material Parameters of a Gas Diffusion Layer Using Tomography Images. *Journal of Fuel Cell Science and Technology*, 5(2): 021006–1 – 021006–9, **2008**.

[15] BERKOWITZ, B. and HANSEN, D. P. A Numerical Study of the Distribution of Water in Partially Saturated Porous Rock. *Transport in Porous Media*, 45: 303–319, **2001**.

[16] BROOKS, R. H. and COREY, A. T. Hydraulic properties of porous media. *Hydrology paper*, 3: 22–27, **1964**.

[17] BROWN, H. W. Capillary Pressure Investigations. *Transactions of the AIME*, 192: 67–74, **1951**.

[18] BRUCE, W. A. and WELGE, H. J. Restored state method for determination of oil in place and connate water. *Oil and Gas Journal*, 46: 223, **1947**.

[19] BURDINE, N. T. Relative permeability calculation from Pore Size Distribution Data. *Transactions of the AIME*, 198: 71–78, **1953**.

[20] CASTON, T. B., MURPHY, A. R. and HARRIS, T. A. L. Effect of weave tightness and structure on the in-plane and through-plane air permeability of woven carbon fibers for gas diffusion layers. *Journal of Power Sources*, 196: 709–716, **2011**.

[21] CERNY, V. Thermodynamical approach to the traveling salesman problem: An efficient simulation algorithm. *Journal of Optimization Theory and Application*, 45(1): 41–51, **1985**.

[22] CHAPUIS, O., PRAT, M., QUINTARD, M., CHANE-KANE, E., GUILLOT, O. and MAYER, N. Two-phase flow and evaporation in model fibrous media: Application to the gas diffusion layer of PEM fuel cells. *Journal of Power Sources*, 178: 258–268, **2008**.

[23] CHEUNG, P., FAIRWEATHER, J. D. and SCHWARTZ, D. T. Characterization of internal wetting in polymer electrolyte membrane gas diffusion layers. *Journal of Power Sources*, 187: 487–492, **2009**.

[24] DAS, P. K., LI, X. and LIU, Z.-S. Effective transport coefficients in PEM fuel cell catalyst and gas diffusion layers: Beyond Bruggeman approximation. *Applied Energy*, 87: 2785–2796, **2010**.

[25] DIN. 18130-1: Baugrund, Untersuchung von Bodenproben; Bestimmung des Wasserdurchlässigkeitsbeiwerts - Teil 1: Laborversuche. **1998**.

[26] DULLIEN, F. A. L. Porous Media - Fluid Transport and Pore Structure (Academic Press, Inc.), **1992**.

[27] DuMu^X. Free and Open-Source Simulator for Flow and Transport Processes in Porous Media. http://dumux.org, **2011**. [accessed 21.08.2011].

[28] DUNE. Distributed and Unified Numerics Environment, modular toolbox for solving PDE. http://www.dune-project.org, **2011**. [accessed 21.08.2011].

[29] DWENGER, S., EIGENBERGER, G. and NIEKEN, U. Measurement of Capillary Pressure-Saturation Relationships Under Defined Compression Levels for Gas Diffusion Media of PEM Fuel Cells. *Transport in Porous Media*, 91(1): 281–294, **2012**.

[30] EDGAR, T. F., HIMMELBLAU, D. M. and LASDON, L. S. Optimization of chemical processes (McGraw-Hill), **2001**.

[31] FAIRWEATHER, J. D., CHEUNG, P., ST-PIERRE, J. and SCHWARTZ, D. T. A microfluidic approach for measuring capillary pressure in PEMFC gas diffusion layers. *Electrochemistry Communications*, 9: 2340–2345, **2007**.

[32] FESER, J., PRASAD, A. and ADVANI, S. Experimental characterization of in-plane permeability of gas diffusion layers. *Journal of Power Sources*, 162: 1226–1231, **2006**.

[33] FLÜCKIGER, R., FREUNBERGER, S. A., KRAMER, D., WOKAUN, A., SCHERER, G. G. and BÜCHI, F. N. Anistropic, effective diffusivity of porous gas diffusion layer materials for PEFC. *Electrochimica Acta*, 54: 551–559, **2008**.

[34] FULLER, E. N., ENSLEY, K. and GIDDINGS, J. C. Diffusion of halogenated hydrocarbons in helium. The effect of structure on collision cross sections. *Journal of Physical Chemistry*, 73(11): 3679–3685, **1969**.

[35] FULLER, E. N. and GIDDINGS, J. C. A Comparison of Methods for Predicting Gaseous Diffusion Coefficients. *Journal of Chromatographic Science*, 3(7): 222–227, **1965**.

[36] FULLER, E. N., SCHETTLER, P. D. and GIDDINGS, J. C. New method for prediction of binar gas-phase diffusion coefficients. *Industrial & Engineering Chemical Research*, 58(5): 18–27, **1966**.

[37] VAN GENUCHTEN, M. T. A Closed-form Equation for Predicting the Hydraulic Conductivity of Unsaturated Soils. *Soil Science Society of America Journal*, 44: 892–898, **1980**.

[38] GOSTICK, J. T., FOWLER, M. W., IOANNIDIS, M. A., PRITZKER, M. D., VOLFKOVICH, Y. M. and SAKARS, A. Capillary pressure and hydrophilic porosity in gas diffusion layers for polymer electrolyte fuel cells. *Journal of Power Sources*, 156: 375–387, **2006**.

[39] GOSTICK, J. T., FOWLER, M. W., PRITZKER, M. D., IOANNIDIS, M. A. and
BEHRA, L. M. In-plane and through-plane gas permeability of carbon fiber
electrode backing layers. *Journal of Power Sources*, 162: 228–238, **2006**.

[40] GOSTICK, J. T., IOANNIDIS, M. A., FOWLER, M. W. and PRITZKER, M. D.
Pore network modeling of fibrous gas diffusion layers for polymer electrolyte
membrane fuel cells. *Journal of Power Sources*, 173: 277–290, **2007**.

[41] GOSTICK, J. T., IOANNIDIS, M. A., FOWLER, M. W. and PRITZKER, M. D.
Direct Measurement of the capillary pressure characteristics of water-air-gas diffusion layer systems for PEM fuel cells. *Electrochemistry Communications*, 10:
1520–1523, **2008**.

[42] GOSTICK, J. T., IOANNIDIS, M. A., FOWLER, M. W. and PRITZKER, M. D.
Wettability and capillary behavior of fibrous gas diffusion media for polymer
electrolyte membrane fuel cells. *Journal of Power Sources*, 194: 433–444, **2009**.

[43] GREENING, D. R. Parallel simulated annealing techniques. *Physica D: Nonlinear
Phenomena*, 42(1-3): 293 – 306, **1990**.

[44] HARKNESS, I., HUSSAIN, N., SMITH, L. and SHARMAN, J. The use of novel
water porosimeter to predict the water handling behaviour of gas diffusion media
used in polymer electrode fuel cells. *Journal of Power Sources*, 193: 122–129,
2009.

[45] HE, G., ZHAO, Z., MING, P., ABULITI, A. and YIN, C. A fractal model
for predicting permeability and liquid water relative permeabilitiy in the gas
diffusion layer (GDL) of PEMFCs. *Journal of Power Sources*, 163: 846–852,
2007.

[46] HELMIG, R. Multiphase Flow and Transport Processes in the Subsurface: A
Contribution to the Modeling of Hydrosystems (Springer), **1997**.

[47] HELMY, A. K., FERREIRO, E. A. and DE BUSSETTI, S. G. The water/graphitic-
carbon interaction energy. *Applied Surface Science*, 253(11): 4966–4969, **2007**.

[48] HUBER, M. personal communication. **2010-2011**.

[49] HUBER, M., SÄCKEL, W., HIRSCHLER, M., HASSANIZADEH, M. and NIEKEN, U. Modeling the dynamics of partial wetting. In Particles - III. International Conference of Particle-Based Methods. Fundamentals and Applications. **2013**.

[50] ISRAELACHVILI, J. Intermolecular And Surface Forces (Academic Press), **2007**.

[51] JIANG, F. and SOUSA, A. C. M. Smoothed Particle Hydrodynamics Modeling of Transverse Flow in Radomly Aligned Fibrous Porous Media. *Transport in Porous Media*, 75: 17–33, **2008**.

[52] KANDLIKAR, S. G., LU, Z., COOKE, D. and DAINO, M. Uneven gas diffusion layer intrusion in gas channel arrays of proton exchange membrane fuel cell and its effects on flow distribution. *Journal of Power Sources*, 194: 328–337, **2009**.

[53] KEIL, F. Diffusion und Chemische Reaktionen in der Gas/Feststoff-Katalyse (Springer), **1999**.

[54] KIRKPATRICK, S., GELATT JR., C. D. and VECCHI, M. P. Optimization by Simulated Annealing. *Science*, 220(4598): 671–680, **1983**.

[55] KLANTE, D., EGGERS, R., HOLZKNECHT, C. and KABELAC, S. Grenzflächenenergien von Teflon und verschiedenen Flüssigkeiten in einer verdichteten Stickstoffatmosphäre. *Forschung im Chemieingenieurwesen*, 67(2): 56–71, **2002**.

[56] KNIGHT, R., CHAPMAN, A. and KNOLL, M. Numerical modeling of microscopic fluid distribution in porous media. *Journal of Applied Physics*, 68(3): 994–1001, **1990**.

[57] KOIDO, T., FURUSAWA, T. and MORIYAMA, K. An approach to modeling two-phase transport in the gas diffusion layer of a proton exchange membrane fuel cell. *Journal of Power Sources*, 175(1): 127–136, **2008**.

[58] LAMANNA, J. M. and KANDLIKAR, S. G. Determination of effective water vapor diffusion coefficient in pemfc gas diffusion layers. *International Journal of Hydrogen Energy*, 36: 5021–5029, **2011**.

[59] LARMINIE, J. and DICKS, A. Fuel Cell Systems Explained (John Wiley & Sons), **2003**.

[60] LAUSER, A. Implementation and Application of a Hysteresis Model for Multiphase Flow and Transport in Porous Media. Master's thesis, University of Stuttgart, **2008**.

[61] LAUSER, A. personal communication. **2010-2011**.

[62] LEE, F.-H. A. Parallel Simulated Annealing on a Message-Passing Multicomputer. Ph.D. thesis, Utah State University, **1995**.

[63] LEVERETT, M. C. Capillary behavior in Porous Solids. *Transactions of the AIME*, 142: 152–169, **1941**.

[64] LIFSHIN, E. X-ray Characterization of Materials (Wiley-VCH), **1999**.

[65] MEIER, F. Stofftransport in Polymerelektrolyt-Membranen für Brennstoffzellen - experimentelle Untersuchung, Modellierung und Simulation. Ph.D. thesis, University of Stuttgart, **2004**.

[66] METROPOLIS, N., ROSENBLUTH, A. W., ROSENBLUTH, M. N., TELLER, A. H. and TELLER, E. Equation of State Calculations by Fast Computing Machines. *Journal of Chemical Physics*, 21(6): 1087–1092, **1953**.

[67] MOHANTY, S. Effect of multiphase fluid saturation on the thermal conductivity if geological media. *Journal of Physics D: Applied Physics*, 30(24): L80–L84, **1997**.

[68] MORSE, R., TERWILLIGER, P. L. and YUSTER, S. T. Relative Permeability Measurements on Small Core Samples. *The Producers Monthly*, 109: 19–25, **1947**.

[69] MPI. Message Passing Interface. http://www.mpi-forum.org, **2011**. [accessed 13.05.2011].

[70] NAM, J. H. and KAVIANY, M. Effective diffusivity and water-saturation distribution in single- and two-layer PEMFC diffusion medium. *International Journal of Heat and Mass Transfer*, 46(24): 4595–4611, **2003**.

[71] NGUYEN, T. V., LIN, G., OHN, H. and WANG, X. Measurement of Capillary Pressure Property of Gas Diffusion Media Used in Proton Exchange Membrane Fuel Cells. *Electrochemical and Solid-State Letters*, 11(8): B127–B131, **2008**.

[72] NUPUS. Non-linearities and Upscaling in Porus Media (international research training group). http://www.nupus.uni-stuttgart.de, **2011**. [accessed 21.08.2011].

[73] OCHS, S. O. Development of a multiphase multicomponent model for PEMFC. *Technical report*, International Research Training Group NUPUS (Nonlinearities and upscaling in porous media), **2008**.

[74] O'HAYRE, R., COLELLA, W., CHA, S.-W. and PRINZ, F. B. Fuel Cell Fundamentals (John Wiley & Sons), **2006**.

[75] OPENMP. An API for multi-platform shared-memory parallel programming in C/C++ and Fortran. http://www.openmp.org, **2011**. [accessed 13.05.2011].

[76] OPEN∇FOAM. The Open Source Computational Fluid Dynamics Toolbox. http://www.openfoam.com, **2011**. [accessed 18.08.2011].

[77] OSOBA, J. S., RICHARDSON, J. G., KERVER, J. K., HAFFORD, J. A. and BLAIR, P. M. Laboratory measurement of relative permeability. *Petroleum Transactions, AIME*, 192: 47–56, **1951**.

[78] OSTADI, H., RAMA, P., LIU, Y., CHEN, R., ZHANG, X. and JIANG, K. 3D reconstruction of a gas diffusion layer and a microporous layer. *Journal of Membrane Science*, 351: 69–74, **2010**.

[79] OTSU, N. A threshold selection method from grey level histograms. *IEEE Transactions on Systems, Man and Cybernetics*, 9(1): 62–66, **1979**.

[80] PARAVIEW. Open Source Scientific Visualization. http://www.paraview.org, **2011**. [accessed 23.07.2011].

[81] PARK, I.-S., DO, D. D. and RODRIGUES, A. E. Measurement of the Effective Diffusivity in Porous Media by the Diffusion Cell Method. *Catalysis Reviews, Science and Engineering*, 38(2): 189–247, **1996**.

[82] PARKER, J. C., LENHARD, R. J. and KUPPUSAMY, T. A parametric model for constitutive properties governing multiphase flow in porous media. *Water Resources Research*, 23(4): 618–624, **1987**.

[83] PASAOGULLARI, U. and WANG, C. Y. Liquid Water Transport in Gas Diffusion Layer of Polymer Electrolyte Fuel Cells. *Journal of The Electrochemical Society*, 151: A399–A406, **2004**.

[84] PURCELL, W. Capillary pressures - their measurement using Mercury and the calculation of permeability therefrom. *Petroleum Transactions, AIME*, 186: 39–48, **1949**.

[85] SCHEIDEGGER, A. E. The Physics of Flow through Porous Media (University of Toronto Press), **1974**.

[86] SERRA, J. Image Analysis and Mathematical Morphology (Academic Press Inc), **1983**.

[87] SHIA, Y., XIAO, J., PANA, M. and YUANA, R. A fractal permeability model for the gas diffusion layer of PEM fuel cells. *Journal of Power Sources*, 160(1): 277–283, **2006**.

[88] SILVERSTEIN, D. L. and FORT, T. Incorporating Low Hydraulic Conductivity in a Numerical Model for Predicting Air-Water Interfacial Area in Wet Unsaturated Particulate Porous Media. *Langmuir*, 16(2): 835–838, **2000**.

[89] SILVERSTEIN, D. L. and FORT, T. Prediction of Air-Water Interfacial Area in Wet Unsaturated Porous Media. *Langmuir*, 16(2): 829–834, **2000**.

[90] SILVERSTEIN, D. L. and FORT, T. Prediction of Water Configuration in Wet Unsaturated Porous Media. *Langmuir*, 16(2): 839–844, **2000**.

[91] SOILLE, P. Morphological Image Analysis: Principles and Applications (Springer), **2003**.

[92] STEPHAN, P., SCHABER, K., STEPHAN, K. and MAYINGER, F. Thermodynamik (Springer), **2010**.

[93] SUKOP, M. C., HUANG, H., LIN, C. L., DEO, M. D., KYEONGSEOK, O. and MILLER, J. D. Distribution of multiphase fluids in porous media: Comparison between lattice Boltzmann modeling and micro-x-ray tomography. *Physical Review E*, 77: 026710, **2008**.

[94] SUNAKAWA, D., OYAMA, S., ARAKI, T. and ONDA, K. Measurement of Diffusion Coefficient and Electro-osmotic Coefficient of Water at PEFC. *Electrochemistry*, 74: 732–736, **2006**.

[95] VTK. Visualization Toolkit. http://www.vtk.org, **2011**. [accessed: 23.07.2011].

[96] WANG. An Alternative Description of Viscous Coupling in Two-phase Flow through Porous Media. *Transport in Porous Media*, 28: 205–219, **1997**.

[97] WANG, Z., WANG, C. Y. and CHEN, K. S. Two-phase flow and transport in the air cathode of proton exchange membrane fuel cells. *Journal of Power Sources*, 94: 40–50, **2001**.

[98] WELGE, H. J. A Simplified Method for Computing Oil Recovery by Gas or Water Drive. *Transactions of the American Institute of Mining*, 195: 91–98, **1952**.

[99] WIESER, M. Stromdichteverteilung und Leistungsverhalten der Polymerelektrolyt-Brennstoffzelle. Ph.D. thesis, University of Stuttgart, **2001**.

[100] WITTE, E., CHAMBERLAIN, R. and FRANKLIN, M. Parallel Simulated Annealing using Speculative Computation. *IEEE Transactions on Parallel and Distributed Systems*, 2: 483–494, **1991**.

[101] WU, R., LIAO, Q., ZHU, X. and WANG, H. A fractal model for determining oxygen effective diffusivity of gas diffusion layer under the dry and wet conditions. *International Journal of Heat and Mass Transfer*, 54(19-20): 4341–4348, **2011**.

[102] WU, R., ZHU, X., LIAO, Q., WANG, H., DING, Y.-D., LI, J. and YE, D.-D. Determination of oxygen effective diffusivity in porous gas diffusion layer using

a three-dimensional pore network model. *Electrochimica Acta*, 55: 7394–7403, **2010**.

[103] WÖHR, M. Instationäres, thermodynamisches Verhalten der Polymermembran-Brennstoffzelle. Ph.D. thesis, University of Stuttgart, **1999**.

[104] ZAMEL, N., ASTRATH, N. G., LI, X., SHEN, J., ZHOU, J., ASTRATH, F. B., WANG, H. and LIU, Z.-S. Experimental measurements of effective diffusion coefficient of oxygen-nitrogen mixture in PEM fuel cell diffusion media. *Chemical Engineering Science*, 65: 931–937, **2010**.

[105] ZAWODZINSKI, T. A., DEROUIN, C., RADZINSKI, S., SHERMAN, R. J., SMITH, V. T., SPRINGER, T. E. and GOTTESFELD, S. Water Uptake by and Transport Through Nafion 117 Membranes. *Journal of the Electrochemical Society*, 140(4): 1041–1047, **2003**.

[106] ZIEGLER, C. and GERTEISEN, D. Validity of two-phase polymer electrolyte membrane fuel cell models with respect to the gas diffusion layer. *Journal of Power Sources*, 188: 184–191, **2009**.